BEAUTIFUL BIOMES

TUNDRA BIOME

by Elizabeth Andrews

Cody Koala
An Imprint of Pop!
popbooksonline.com

abdobooks.com
Published by Pop!, a division of ABDO, PO Box 398166, Minneapolis, Minnesota 55439.

Printed in the United States of America, North Mankato, Minnesota

102021
012022

THIS BOOK CONTAINS RECYCLED MATERIALS

Cover Photo: Shutterstock Images
Interior Photos: Shutterstock Images, 1, 5 (top) (bottom left) (bottom center), 7, 8, 9, 13, 14–15, 17 (top) (bottom left) (bottom center), 18; AscentXmedia/Getty Images, 10–11; Bosca78/Getty Images, 20–21

Editor: Tyler Gieseke
Series Designer: Laura Graphenteen

Library of Congress Control Number: 2021942295

Publisher's Cataloging-in-Publication Data
Names: Andrews, Elizabeth, author.
Title: Tundra biome / by Elizabeth Andrews
Description: Minneapolis, Minnesota : Pop!, 2022 | Series: Beautiful biomes | Includes online resources and index.
Identifiers: ISBN 9781098241056 (lib. bdg.) | ISBN 9781098241759 (ebook)
Subjects: LCSH: Tundras--Juvenile literature. | Biotic communities--Juvenile literature. | Habitats--Juvenile literature. | Life zones--Juvenile literature. | Tundra animals--Juvenile literature. | Tundra plants--Juvenile literature. | Tundra ecology--Juvenile literature.
Classification: DDC 577.5--dc23

Hello! My name is

Cody Koala

Pop open this book and you'll find QR codes like this one, loaded with information, so you can learn even more!

Scan this code* and others like it while you read, or visit the website below to make this book pop.

popbooksonline.com/tundra-biome

*Scanning QR codes requires a web-enabled smart device with a QR code reader app and a camera.

Table of Contents

Chapter 1

The Coldest Biome

A biome is a large, natural area. It is known for the plants and animals that live there, and its **climate**.

tundra

Watch a video here!

forest

freshwater

Chapter 2

Tundra Biomes

The tundra biome is usually a cold and icy place. There are two kinds of tundra. Arctic tundras are far north in the **Arctic Circle**. The winters there are very long. **Temperatures** stay around -30°F (-34.4°C).

Learn more here!

Summers in the Arctic are short. They can get up to 50°F (10°C). All plant growth happens during the short

summer months. Animal populations grow too. That's when babies are born.

Alpine tundras are found high in the mountains. They can occur in any mountain range that gets tall enough.

It is cold and windy up there. Temperatures are usually 10°F (-12.2°C) in the winter and 50°F (10°C) in the summer.

Temperatures go down when air pressure decreases. Air pressure lowers high up in the mountains, so it gets colder there.

Chapter 3

Tundra Plants

It is hard for plants to grow in tundras. The **climate** is too cold. Wind blows too hard for trees to survive. The ground never **thaws** enough to support large plants with deep roots.

NO TREES IN THE TUNDRA

Learn more here!

Most plants in tundras grow close to the ground. Shrubs, mosses, grasses, and flowers have short roots.

The summer sun shines all day and night in the Arctic tundra. This helps plants grow quickly.

They grow quickly in the summer. Flowers bloom for only a few days.

Chapter 4

Tundra Animals

Tundra animals have many **adaptions**. Some animals' fur changes color with the seasons. In winter they are white like the snow. In summer they are darker like the ground.

Complete an activity here!

Arctic ground squirrels survive the cold winter by **hibernating**. Eight months can feel like 12 days to them during hibernation.

There are tundra animals of all shapes and sizes. Small rodents like Arctic ground squirrels and lemmings burrow in the ground. Giant polar bears live near the ocean. They hunt seals beneath the water.

Alpine tundras have different kinds of animals. These animals are built to live high up in the mountains. Musk ox and mountain goats have hooves. They are good climbers.

Making Connections

Text-to-Self

Which tundra animal do you think has the best adaptions for the cold climate?

Text-to-Text

Have you read a book about any other biomes? If so, how are they similar to and different from tundras?

Text-to-World

There are groups of people that live in the tundra. Do you think life is easy for them? What challenges would they face?

Glossary

adaption – a change in a plant or animal that helps it survive in a certain place.

Arctic Circle – an imaginary line drawn around the top of the earth that marks the boundary of the Arctic.

climate – weather conditions that are usual in an area over a long period of time.

hibernate – to spend a period of time, such as the winter, in deep sleep.

temperature – a measure of hotness or coldness.

thaw – to go from being a frozen solid to a soft solid.

Index

Online Resources

popbooksonline.com

Thanks for reading this Cody Koala book!

Scan this code* and others like it in this book, or visit the website below to make this book pop!

popbooksonline.com/tundra-biome

*Scanning QR codes requires a web-enabled smart device with a QR code reader app and a camera.

BEAUTIFUL BIOMES

MARINE BIOME

by Elizabeth Andrews

Cody Koala
An Imprint of Pop!
popbooksonline.com

abdobooks.com
Published by Pop!, a division of ABDO, PO Box 398166, Minneapolis, Minnesota 55439.

Printed in the United States of America, North Mankato, Minnesota

102021
012022

THIS BOOK CONTAINS RECYCLED MATERIALS

Cover Photo: Shutterstock Images
Interior Photos: Shutterstock Images, 1, 5 (top) (bottom left) (bottom center), 6–7, 9, 10–11, 12, 13, 15, 16 (manatee) (mangrove), 19, 20, 21 (top) (bottom)

Editor: Tyler Gieseke
Series Designer: Laura Graphenteen

Library of Congress Control Number: 2021942283

Publisher's Cataloging-in-Publication Data
Names: Andrews, Elizabeth, author.
Title: Marine biome / by Elizabeth Andrews
Description: Minneapolis, Minnesota : Pop!, 2022 | Series: Beautiful biomes | Includes online resources and index.
Identifiers: ISBN 9781098241049 (lib. bdg.) | ISBN 9781098241742 (ebook)
Subjects: LCSH: Ocean--Juvenile literature. | Biotic communities--Juvenile literature. | Habitats--Juvenile literature. | Life zones--Juvenile literature. | Marine animals--Juvenile literature. | Marine plants--Juvenile literature. | Marine ecology--Juvenile literature.
Classification: DDC 577.7--dc23

Hello! My name is

Cody Koala

Pop open this book and you'll find QR codes like this one, loaded with information, so you can learn even more!

Scan this code* and others like it while you read, or visit the website below to make this book pop.
popbooksonline.com/marine-biome

*Scanning QR codes requires a web-enabled smart device with a QR code reader app and a camera.

Table of Contents

Chapter 1

Under the Sea

A biome is a large, natural area. It is known for the plants and animals that live there, and its **climate**.

marine

desert

tundra

Watch a video here!

Marine water biomes cover most of the earth. They provide rainwater for land. Plants living in these biomes make lots of oxygen for the planet. Marine biomes come in three types.

Chapter 2

Three Marine Biomes

The ocean is a large body of water. The water is salty. Its temperature ranges greatly all over the world. The ocean near the **equator** is warmer than the ocean near the north and south **poles**.

Learn more here!

Coral reefs are found in warm, shallow parts of the

ocean. They work as **barriers** along some **continents**.

Estuaries are special. They occur where freshwater rivers or streams flow into the ocean. Some fish and

waterbirds go to estuaries to have babies. Those biomes are also safe places for babies to grow.

Chapter 3

Marine Plants

Plants grow in all marine biomes. Plants need sunlight to grow. Some places in the ocean are too deep for light to reach. Plants won't grow there. Near the shore, kelp and algae grow.

Learn more here!

THE MANGROVE

Mangrove roots stick up from the mud. They have to be very strong, so they don't wash away with the waves. All kinds of animals live among the roots.

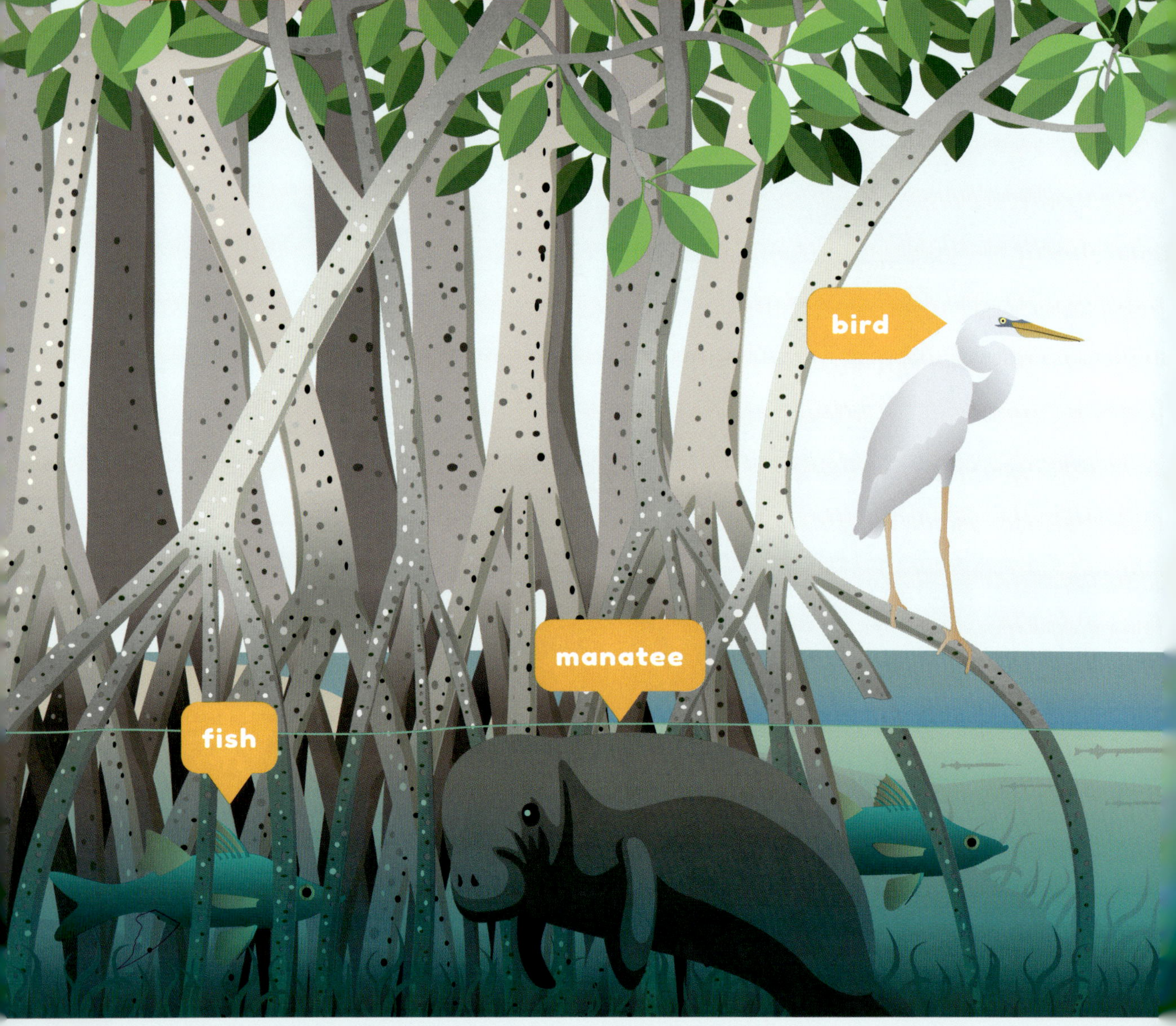

Seagrass meadows and mangrove forests grow in certain marine biomes. They offer food and shelter to many animals.

Manatees and sea turtles eat the seagrass in marine meadows.

Chapter 4

Marine Animals

Many animals live in marine biomes. Coral is sometimes mistaken for a plant. But it is actually an animal. Many grow together to make coral reefs. Starfish and sea turtles live there.

Complete an activity here!

Swimming **mammals** like whales and dolphins live in the ocean. They go to the surface to get the oxygen they need. Fish can breathe oxygen underwater.

The marine biome supports the largest animal on Earth. It is the blue whale. It can weigh as much as 33 elephants!

Birds like herons and ducks make their homes in estuaries. **Crustaceans**, starfish, and clams live in tide pools along some rocky ocean shores.

Making Connections

Text-to-Self

Have you ever seen a marine biome? If you have, which type was it? If you haven't, what type would you most want to see?

Text-to-Text

Have you read a book about any other biomes? If so, how are they similar to and different from marine biomes?

Text-to-World

Some parts of the ocean are struggling because of climate change. Do you know of any ways people are trying to help marine biomes?

Glossary

barrier – a natural feature that blocks or prevents movement, like ocean waves.

climate – weather conditions that are usual in an area over a long period of time.

continents – seven large pieces of land on the globe.

crustacean – any of a group of animals with a hard shell and jointed legs.

equator – an imaginary line around the middle of Earth, halfway between the north and south poles.

mammal – an animal that makes milk to feed its young and usually has hair or fur on its skin.

pole – either end of Earth. The north pole and the south pole are opposite each other.

Index

Online Resources

popbooksonline.com

Thanks for reading this Cody Koala book!

Scan this code* and others like it in this book, or visit the website below to make this book pop!

popbooksonline.com/marine-biome

*Scanning QR codes requires a web-enabled smart device with a QR code reader app and a camera.

BEAUTIFUL BIOMES

GRASSLAND BIOME

by Elizabeth Andrews

Cody Koala
An Imprint of Pop!
popbooksonline.com

abdobooks.com
Published by Pop!, a division of ABDO, PO Box 398166, Minneapolis, Minnesota 55439.

Printed in the United States of America, North Mankato, Minnesota

102021
012022

THIS BOOK CONTAINS RECYCLED MATERIALS

Cover Photo: Lubo Ivanko/Getty Images
Interior Photos: Shutterstock Images, 1, 5 (top, bottom left), 6–7, 9, 10, 13, 14, 16–17, 18–19, 20; deimagine/Getty Images, 5 (bottom center)

Editor: Tyler Gieseke
Series Designer: Laura Graphenteen

Library of Congress Control Number: 2021942246

Publisher's Cataloging-in-Publication Data
Names: Andrews, Elizabeth, author.
Title: Grassland biome / by Elizabeth Andrews
Description: Minneapolis, Minnesota : Pop!, 2022 | Series: Beautiful biomes | Includes online resources and index.
Identifiers: ISBN 9781098241032 (lib. bdg.) | ISBN 9781098241735 (ebook)
Subjects: LCSH: Grasslands--Juvenile literature. | Biotic communities--Juvenile literature. | Habitats--Juvenile literature. | Life zones--Juvenile literature. | Grassland animals--Juvenile literature. | Grassland plants--Juvenile literature. | Grassland ecology--Juvenile literature.
Classification: DDC 577.4--dc23

Hello! My name is

Cody Koala

Pop open this book and you'll find QR codes like this one, loaded with information, so you can learn even more!

Scan this code* and others like it while you read, or visit the website below to make this book pop.

popbooksonline.com/grassland-biome

*Scanning QR codes requires a web-enabled smart device with a QR code reader app and a camera.

Table of Contents

Chapter 1

Grassy Plains

A biome is a large, natural area. It is known for the plants and animals that live there, and its **climate**.

grassland

desert

freshwater

Watch a video here!

A plain is a large, flat area with few trees.

Grasslands are biomes that are covered mostly by grasses. They get the perfect amount of rain for

these plants to grow. There is not enough **precipitation** for larger plants like trees to grow.

Chapter 2

Types of Grasslands

Grasslands grow all around the world. There are two main types. Tropical grasslands called savannas grow near the **equator**. They are warm and have a rainy and a dry season.

Learn more here!

Temperate grasslands are farther away from the equator. They have some of the world's best soils. Cold winters provide most of the yearly **precipitation** as snow. The summers are hot.

Chapter 3

Grassland Plants

Forests cannot grow on grasslands. There is not enough rain. The most important plants in these biomes are grasses and shrubs. They can grow back quickly after wildfires.

Wildfires help prevent trees from taking over grasslands.

Learn more here!

Elephants stomp down bushes
and trees when they move
around savannas in Africa.

Short and tall grasses like lemon grass grow in savannas. Large, **grazing mammals**, like elephants, keep too many trees from growing. Frequent fires also prevent trees and shrubs from taking over.

Meadow grass and purple needlegrass cover the ground in **temperate**

grasslands. Wildflowers grow there too. They make great food for **pollinators**.

The colors, shapes, and scents of flowers attract pollinators to them.

Chapter 4

Grassland Animals

Savannas are home to many amazing animals. Zebras and antelope move in large groups eating the grasses. They are hunted by lions and cheetahs. Kangaroos live in Australian savannas.

Complete an activity here!

Rodents and insects can burrow underground. This helps them survive any wildfires.

Temperate grasslands have lots of insects. Prairie dogs and gophers make their homes underground. Herds of bison live here. They keep the grasses short with the help of other **grazing** animals.

Making Connections

Text-to-Self

Would you rather see animals from tropical grasslands or temperate grasslands? Why?

Text-to-Text

Have you read a book about any other biomes? If so, how are they similar to and different from grasslands?

Text-to-World

Are wildfires good or bad for grasslands? Do you think the animals that live there are prepared to survive wildfires when they happen?

Glossary

climate – weather conditions that are usual in an area over a long period of time.

equator – an imaginary line around the middle of Earth, halfway between the north and south poles.

graze – to feed on plants like grasses.

mammal – an animal that makes milk to feed its young and usually has hair or fur on its skin.

pollinator – an animal that carries pollen from one part of a flower or plant to another.

precipitation – moisture such as rain, hail, or snow that falls to Earth.

temperate – having mild weather.

Index

Online Resources

popbooksonline.com

Thanks for reading this Cody Koala book!

Scan this code* and others like it in this book, or visit the website below to make this book pop!

popbooksonline.com/grassland-biome

*Scanning QR codes requires a web-enabled smart device with a QR code reader app and a camera.

BEAUTIFUL BIOMES

DESERT BIOME

by Elizabeth Andrews

Cody Koala
An Imprint of Pop!
popbooksonline.com

abdobooks.com
Published by Pop!, a division of ABDO, PO Box 398166, Minneapolis, Minnesota 55439.

Printed in the United States of America, North Mankato, Minnesota

102021
012022

THIS BOOK CONTAINS RECYCLED MATERIALS

Cover Photo: Shutterstock Images
Interior Photos: Shutterstock Images, 1, 5 (top) (bottom left) (bottom center), 6–7, 10, 11, 15 (right plant), 17,19, 20; diegograndi/Getty Images, 9; Erlantz Pérez Rodríguez/Getty Images, 12; skalapendra/Getty Images, 15 (left plant)

Editor: Tyler Gieseke
Series Designer: Laura Graphenteen

Library of Congress Control Number: 2021942251

Publisher's Cataloging-in-Publication Data
Names: Andrews, Elizabeth, author.
Title: Desert biome / by Elizabeth Andrews
Description: Minneapolis, Minnesota : Pop!, 2022 | Series: Beautiful biomes | Includes online resources and index.
Identifiers: ISBN 9781098241001 (lib. bdg.) | ISBN 9781098241704 (ebook)
Subjects: LCSH: Deserts--Juvenile literature. | Biotic communities--Juvenile literature. | Habitats--Juvenile literature. | Life zones--Juvenile literature. | Desert animals--Juvenile literature. | Desert plants--Juvenile literature. | Desert ecology--Juvenile literature.
Classification: DDC 577.54--dc23

Hello! My name is

Cody Koala

Pop open this book and you'll find QR codes like this one, loaded with information, so you can learn even more!

Scan this code* and others like it while you read, or visit the website below to make this book pop.

popbooksonline.com/desert-biome

*Scanning QR codes requires a web-enabled smart device with a QR code reader app and a camera.

Table of Contents

Chapter 1

The Driest Biome

A biome is a large, natural area. It is known for the plants and animals that live there, and its **climate**.

desert

Watch a video here!

freshwater

forest

Deserts are the driest biome. They cover 20 percent of Earth. To be a desert, the amount of water

that dries up must be more
than the amount of rain that
falls each year.

Chapter 2

The Four Deserts

Desert biomes can be broken down into four types. Hot and dry deserts have very warm weather all year. Rain is rare.

Learn more here!

Semiarid deserts are also very hot. But they have seasons. Summers there are long and dry.

The continent of Antarctica is a desert!

Cold deserts have long, freezing winters. They even get some snow.

Coastal deserts sit along oceans. These deserts are more **humid** because fog rolls in from the water. They still get very little rainfall, but seasons are more clear.

Chapter 3

Desert Plants

Desert plants are special. They have **adapted** to the harsh **climate**. Some have short roots that quickly absorb any rain. Other plants have long roots that use water stored deep underground.

Learn more here!

Plants like cacti have thick, waxy skin. This saves water and **reflects** heat. Cacti also have spikes to keep animals from stealing their stored water.

A cactus called prickly pear grows pretty, pink flowers and fruits.

Chapter 4

Desert Animals

Animals that live in the desert have also **adapted**. Many live underground to stay cool. They come out at night to look for food. Most will get all the water they need from their food.

Camel humps store fat. They use it when food is hard to find.

Complete an activity here!

Reptiles and insects can survive well in the desert. Not many **mammals** live there. The ones that do have big ears to hear **prey** underground. They might also have furry paws to protect against the hot sand.

Making Connections

Text-to-Self

What sort of adaptions would you have to make if you lived in a desert?

Text-to-Text

Have you read a book about any other biomes? If so, how are they similar to and different from deserts?

Text-to-World

Do you know anything about the people who live in deserts? What do you think their lives would look like?

Glossary

adapt – to change something so that it is easier to live in a place.

climate – weather conditions that are usual in an area over a long period of time.

humid – having a high amount of moisture in the air.

mammal – an animal that makes milk to feed its young and usually has hair or fur on its skin.

prey – an animal that is hunted and eaten by another animal.

reflect – to throw back light instead of take it in.

Index

Online Resources

popbooksonline.com

Thanks for reading this Cody Koala book!

Scan this code* and others like it in this book, or visit the website below to make this book pop!

popbooksonline.com/desert-biome

*Scanning QR codes requires a web-enabled smart device with a QR code reader app and a camera.

BEAUTIFUL BIOMES

FOREST BIOME

by Elizabeth Andrews

Cody Koala
An Imprint of Pop!
popbooksonline.com

abdobooks.com
Published by Pop!, a division of ABDO, PO Box 398166, Minneapolis, Minnesota 55439.

Printed in the United States of America, North Mankato, Minnesota

102021
012022
THIS BOOK CONTAINS RECYCLED MATERIALS

Cover Photo: Shutterstock Images
Interior Photos: Shutterstock Images, 1, 6, 9, 13, 19; RondaKimbrow/Getty Images, 5 (top); mihtiander/Getty Images, 5 (bottom left); lucky-photographer/Getty Images, 5 (bottom center); Jillian Cooper/Getty Images, 10–11; THEPALMER/Getty Images, 14; Burachet/Getty Images, 15; Mirosław Nowaczyk/Alamy Stock Photo, 16; pablo_arriaran/Getty Images, 20

Editor: Tyler Gieseke
Series Designer: Laura Graphenteen

Library of Congress Control Number: 2021942250

Publisher's Cataloging-in-Publication Data
Names: Andrews, Elizabeth, author.
Title: Forest biome / by Elizabeth Andrews
Description: Minneapolis, Minnesota : Pop!, 2022 | Series: Beautiful biomes | Includes online resources and index.
Identifiers: ISBN 9781098241018 (lib. bdg.) | ISBN 9781098241711 (ebook)
Subjects: LCSH: Forests and forestry--Juvenile literature. | Biotic communities--Juvenile literature. | Habitats--Juvenile literature. | Life zones--Juvenile literature. | Forest animals--Juvenile literature. | Forest plants--Juvenile literature. | Forest ecology--Juvenile literature.
Classification: DDC 577.3--dc23

Hello! My name is

Cody Koala

Pop open this book and you'll find QR codes like this one, loaded with information, so you can learn even more!

Scan this code* and others like it while you read, or visit the website below to make this book pop.

popbooksonline.com/forest-biome

*Scanning QR codes requires a web-enabled smart device with a QR code reader app and a camera.

Table of Contents

Chapter 1

Among the Trees

A biome is a large, natural area. It is known for the plants and animals that live there, and its **climate**.

forest

marine

desert

Watch a video here!

A forest biome is an area covered with trees and woody plants. About one-third of all land is covered by forests. There are three kinds of forests.

Chapter 2

Temperate Forests

Temperate forests have mild **climates**. They are not too cold or too hot. Temperate forests experience four seasons. Leaves fall in autumn and come back in spring.

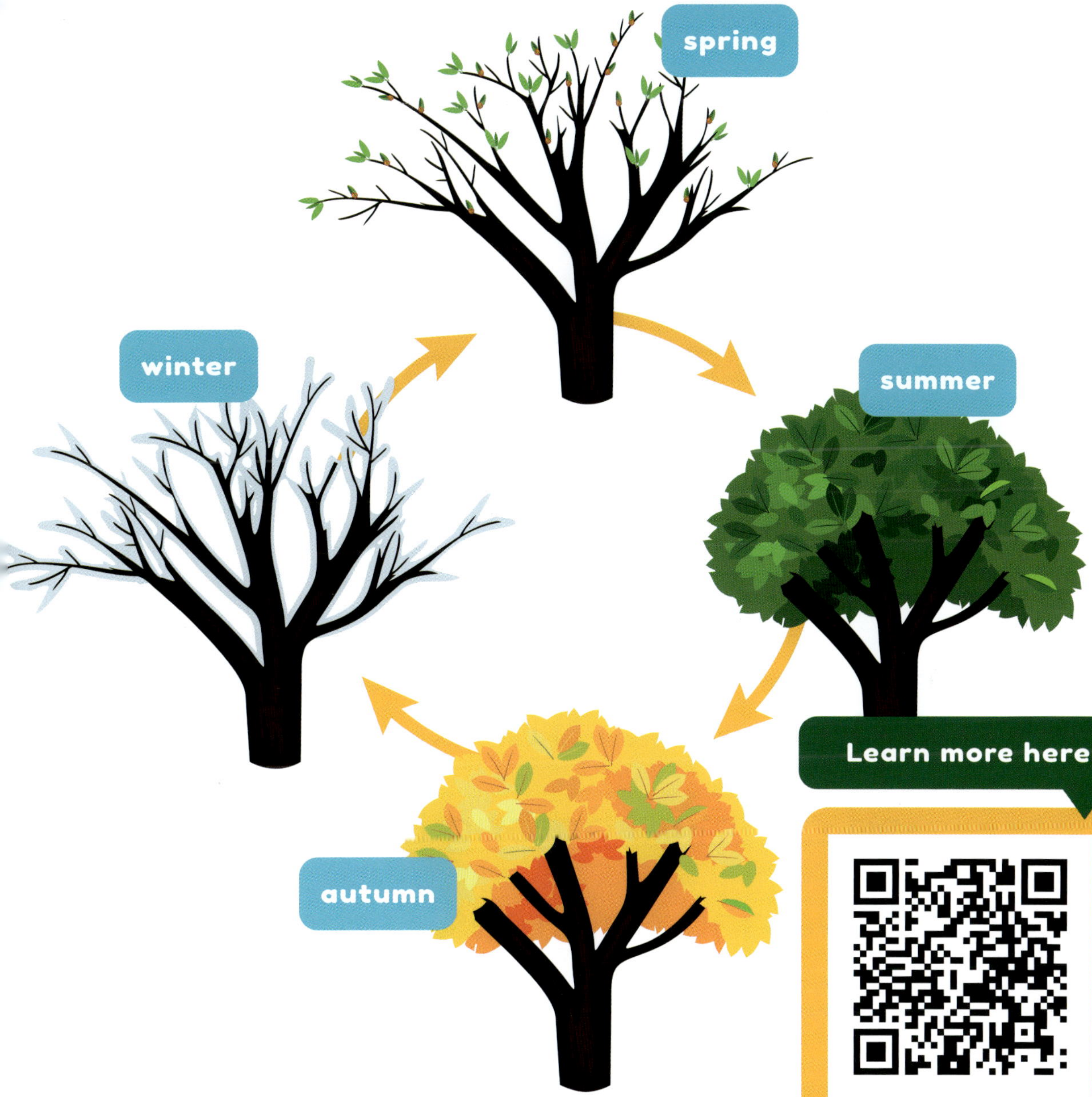

Learn more here!

Animals like black bears and deer live among large maple and oak trees. Ferns and wildflowers grow on the

Black bears eat berries from the bushes that grow in forests.

forest floor. Insects, birds, and rodents also live there. They make their homes in the trees or on the ground.

Chapter 3

Tropical Forests

Tropical forests grow near the **equator**. Tropical forests can receive up to 400 inches (1,000 cm) of rain each year! The weather is hot and **humid** all year long.

Learn more here!

Half of Earth's plant and animal **species** live in tropical forests.

Mangoes, bananas, and papayas grow in tropical forests. Monkeys love these fruits!

Trees grow large due to rain and warm weather. Many of the animals eat fruit! Meat eaters like jaguars and anacondas also live there.

The Wasai tree roots are good for kidney health.

Tropical forests are helpful biomes. They are home to plants that are used to make life-saving medicines. The plants in these biomes also take in **greenhouse gases** that hurt the Earth.

Chapter 4

Boreal Forests

Boreal forests are found far north of the **equator**. Their winters are long, snowy, and cold. Warmer weather in the summer brings the most **precipitation** for these forests.

Boreal forests are also called taigas.

Complete an activity here!

Evergreens are the most common trees. They do not drop their leaves in winter. **Mammals** that can handle the cold **climate**, like bears and moose, live in boreal forests. Woodpeckers and chickadees make homes in the trees.

Making Connections

Text-to-Self

Which kind of forest would you want to live in? Write a few sentences about why you chose this forest.

Text-to-Text

Have you read a book about any other biomes? If so, how are they similar to and different from forests?

Text-to-World

How do you think forests help the world? Is it important to protect them?

Glossary

climate – weather conditions that are usual in an area over a long period of time.

equator – an imaginary line around the middle of Earth, halfway between the North and South poles.

greenhouse gases – gases, such as carbon dioxide, that trap heat in Earth's atmosphere.

humid – having a high amount of moisture in the air.

mammal – an animal that makes milk to feed its young and usually has hair or fur on its skin.

precipitation – moisture such as rain, hail, or snow that falls to Earth.

species – living things that are very much alike.

Index

Online Resources

popbooksonline.com

Thanks for reading this Cody Koala book!

Scan this code* and others like it in this book, or visit the website below to make this book pop!

popbooksonline.com/forest-biome

*Scanning QR codes requires a web-enabled smart device with a QR code reader app and a camera.

BEAUTIFUL BIOMES

FRESHWATER BIOME

by Elizabeth Andrews

Cody Koala
An Imprint of Pop!
popbooksonline.com

abdobooks.com
Published by Pop!, a division of ABDO, PO Box 398166, Minneapolis, Minnesota 55439.

Printed in the United States of America, North Mankato, Minnesota

102021
012022

THIS BOOK CONTAINS RECYCLED MATERIALS

Cover Photo: Shutterstock Images
Interior Photos: Josef Mohyla/Getty Images, 1; deimagine/Getty Images, 5 (top); Getty Images, 5 (bottom left); Ron and Patty Thomas/Getty Images, 5 (bottom center); Shutterstock Images, 6 (watershed), 6 (clouds), 12, 15, 16–17, 19, 20; borchee/Getty Images, 9; Onfokus/Getty Images, 10–11

Editor: Tyler Gieseke
Series Designer: Laura Graphenteen

Library of Congress Control Number: 2021942297

Publisher's Cataloging-in-Publication Data
Names: Andrews, Elizabeth, author.
Title: Freshwater biome / by Elizabeth Andrews
Description: Minneapolis, Minnesota : Pop!, 2022 | Series: Beautiful biomes | Includes online resources and index.
Identifiers: ISBN 9781098241025 (lib. bdg.) | ISBN 9781098241728 (ebook)
Subjects: LCSH: LCSH: Rivers--Juvenile literature. | Biotic communities--Juvenile literature. | Habitats Juvenile literature. | Life zones--Juvenile literature. | Stream animals--Juvenile literature. | Stream plants--Juvenile literature. | Stream ecology--Juvenile literature.
Classification: DDC 577.6--dc23

Hello! My name is

Cody Koala

Pop open this book and you'll find QR codes like this one, loaded with information, so you can learn even more!

Scan this code* and others like it while you read, or visit the website below to make this book pop.
popbooksonline.com/freshwater-biome

*Scanning QR codes requires a web-enabled smart device with a QR code reader app and a camera.

Table of Contents

Chapter 1

Fresh and Flowing

A biome is a large, natural area. It is known for the plants and animals that live there, and its **climate**.

Freshwater biomes are sources of drinking water!

freshwater

marine

desert

Watch a video here!

A WATERSHED
precipitation
creek
river
lake

Freshwater biomes hold only three percent of Earth's water. They are made up of less than one percent salt.

Watersheds are areas that help collect fresh water. Fresh water that falls into a watershed flows to lakes, rivers, or wetlands.

Chapter 2

Freshwater Biomes

Ponds and lakes are **standing** bodies of fresh water. They are all over the world in many shapes and sizes. Water temperatures in ponds and lakes can vary greatly.

Lake Superior is the largest freshwater lake in the world. It covers 31,000 square miles (80,300 km^2)!

Learn more here!

Streams and rivers are freshwater biomes that flow in one direction. A river is cleanest at its **headwaters**. It picks up stuff while it flows.

This makes the water cloudier at the **mouth**. Most animals live near the middle of rivers.

Wetlands can be swamps, marshes, or bogs.

Wetlands are a combination of land and still water. Wetlands can be year-round or seasonal. They help prevent floods. They hold extra water.

Chapter 3

Freshwater Plants

Freshwater biomes are very important. They are home to plants that create oxygen for people and animals to breathe. Many plants that grow in fresh water are food for animals that live there.

Learn more here!

Water lilies have wide leaves that let them float on the surface of water. Cattails grow along the shores of

lakes and ponds. The slimy green stuff that floats on some water is called algae. Ducks like to eat it.

Chapter 4

Freshwater Animals

Freshwater biomes make good homes for animals. Trout and salmon swim in rivers. Giant rock sturgeon swim in the Great Lakes. **Waterfowl** fly to and from wetlands all over the world.

Complete an activity here!

Mammals also live in freshwater biomes. River otters and beavers have webbed feet that make them great swimmers. Beavers even build **dams** in streams. This can make new freshwater biomes!

Making Connections

Text-to-Self

There are three kinds of freshwater biomes. Which would you like to spend the afternoon next to most?

Text-to-Text

Have you read a book about any other biomes? If so, how are they similar to and different from freshwater biomes?

Text-to-World

What do freshwater biomes give to the world? Do you think it is important to protect them?

Glossary

climate – weather conditions that are usual in an area over a long period of time.

dam – something that holds back water.

headwater – the source of a stream or river.

mammal – an animal that makes milk to feed its young and usually has hair or fur on its skin.

mouth – the place where a river enters a large body of water.

standing – not moving.

waterfowl – birds that spend a lot of time in water.

Index

Online Resources

popbooksonline.com

Thanks for reading this Cody Koala book!

Scan this code* and others like it in this book, or visit the website below to make this book pop!

popbooksonline.com/freshwater-biome

*Scanning QR codes requires a web-enabled smart device with a QR code reader app and a camera.

SCIENCE QUESTIONS

HOW DOES PLUMBING WORK?

by Elizabeth Andrews

Cody Koala
An Imprint of Pop!
popbooksonline.com

abdobooks.com
Published by Pop!, a division of ABDO, PO Box 398166, Minneapolis, Minnesota 55439.

Printed in the United States of America, North Mankato, Minnesota

102021
012022

THIS BOOK CONTAINS RECYCLED MATERIALS

Cover Photo: Shutterstock Images
Interior Photos: Getty Images, 6, 9, 13 (top), 16; Shutterstock Images, 5, 7, 10, 13 (bottom row), 14, 15, 17, 19, 20

Editor: Grace Hansen
Series Designer: Laura Graphenteen

Library of Congress Control Number: 2021942299

Publisher's Cataloging-in-Publication Data
Names: Andrews, Elizabeth, author.
Title: How does plumbing work? / by Elizabeth Andrews
Description: Minneapolis, Minnesota : Pop!, 2022 | Series: Science questions | Includes online resources and index.
Identifiers: ISBN 9781098241100 (lib. bdg.) | ISBN 9781098241803 (ebook)
Subjects: LCSH: Plumbing--Juvenile literature. | Water-pipes--Juvenile literature. | Drainage pipes--Juvenile literature. | Children's questions and answers--Juvenile literature.
Classification: DDC 500--dc23

Hello! My name is

Cody Koala

Pop open this book and you'll find QR codes like this one, loaded with information, so you can learn even more!

Scan this code* and others like it while you read, or visit the website below to make this book pop.

popbooksonline.com/plumbing

*Scanning QR codes requires a web-enabled smart device with a QR code reader app and a camera.

Table of Contents

Chapter 1

What Is Plumbing?

Is it magic that makes water fall from a faucet at the twist of a knob? It must be! But it's not magic. It's just plumbing!

Watch a video here!

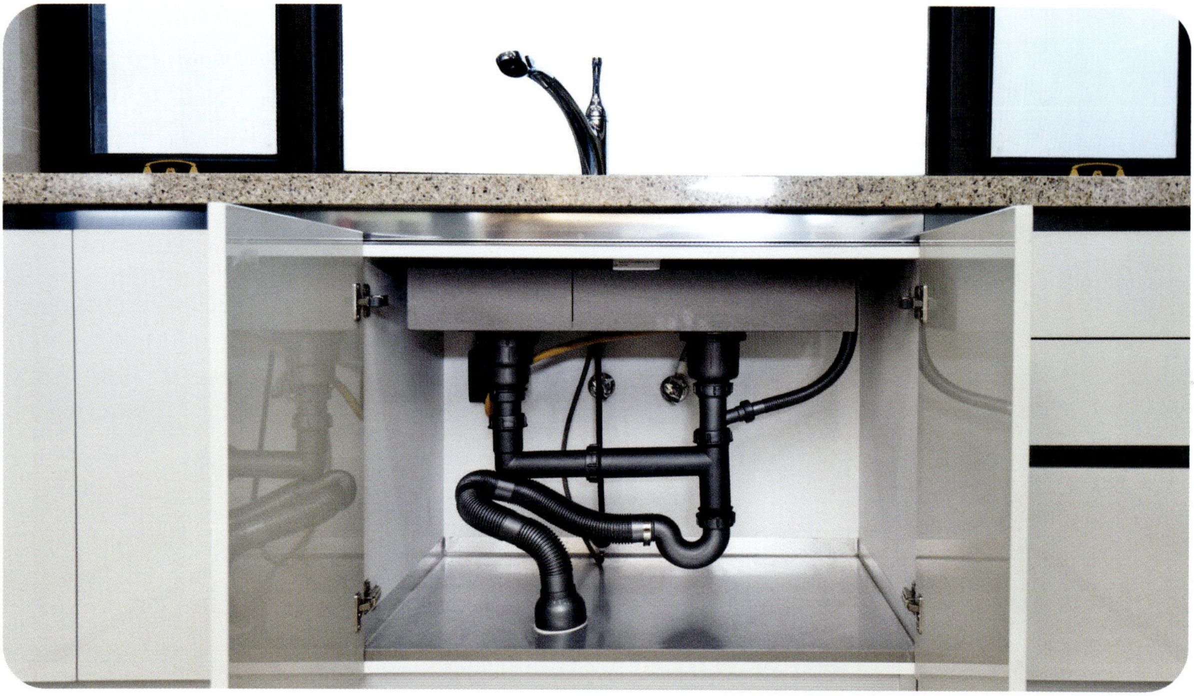

Plumbing has to do with the pipes that bring clean, safe water to homes.

Plumbing also moves waste away from the home. A person who fixes or places plumbing is called a plumber.

Chapter 2

Why Is Plumbing Important?

People need plumbing to stay healthy. In the past people would dump waste into water sources. **Diseases** can grow in unclean water. Many people got sick and died when they used the water.

Learn more here!

In the 1800s people began to understand the importance of clean water. Cities thought about ways to safely move waste out of homes. They also made better **systems** to get clean water into homes.

Sir John Harington invented the first flushing toilet. That's why a nickname for a toilet is "the John!"

Chapter 3

The Two Types

Plumbing today is put together with metal or plastic materials. The water is moved with high-**pressure pumps**.

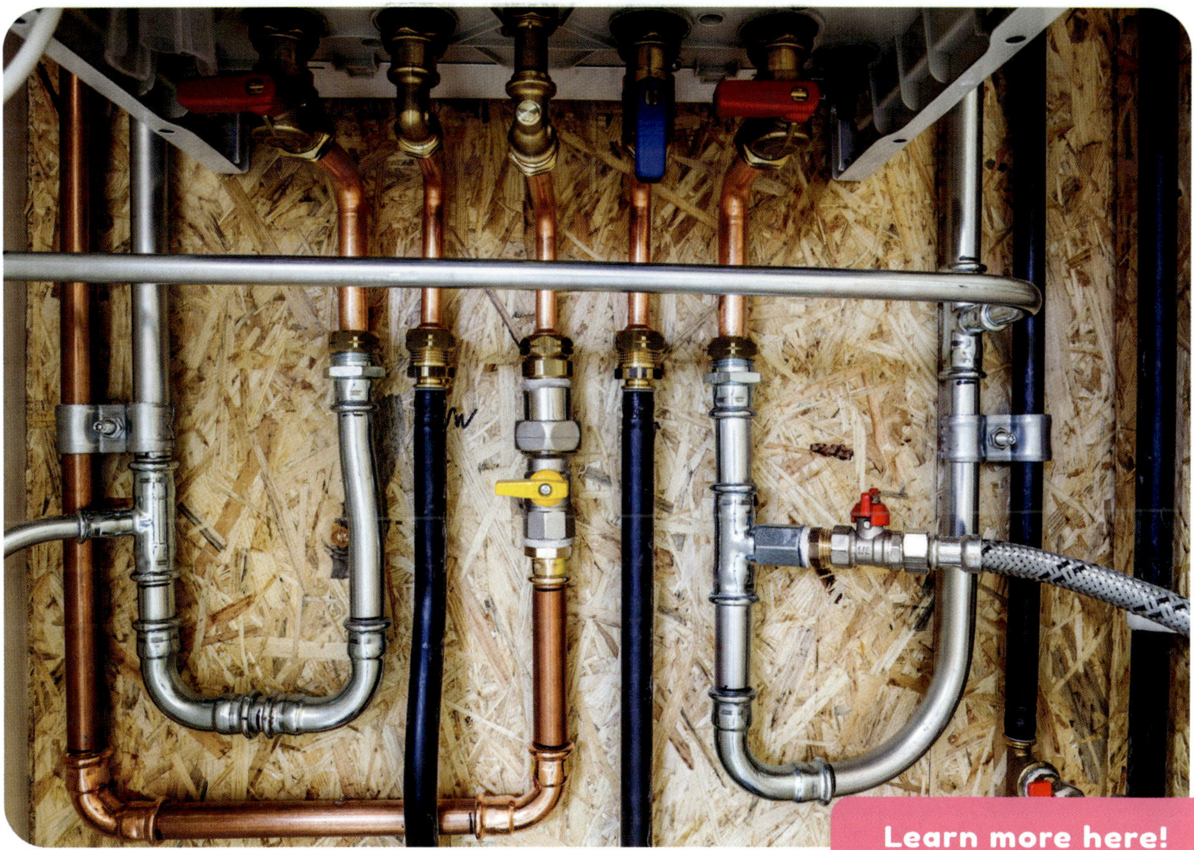

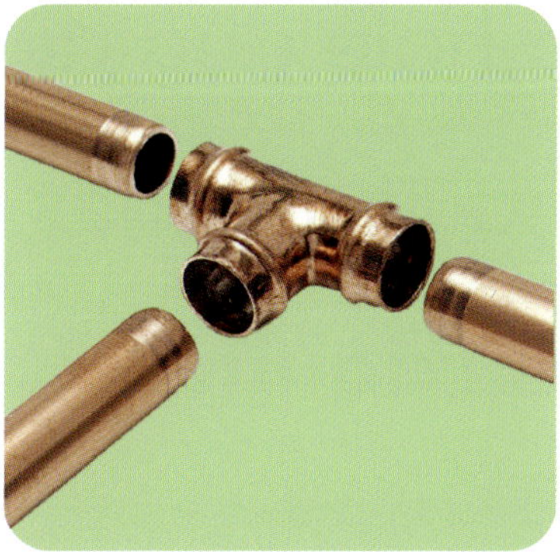

Learn more here!

Well water is used in places far away from cities.

There are two ways to get water to your home. The first way to get water is from a well. A well is a hole dug deep into the ground. It reaches stored water in an **aquifer**.

The water is pumped up into pipes called water mains. These lead to a storage tank. The tank is usually in the basement of a home.

Another way to get water is from the city. Cities store water from lakes and rivers. The water could also come from aquifers.

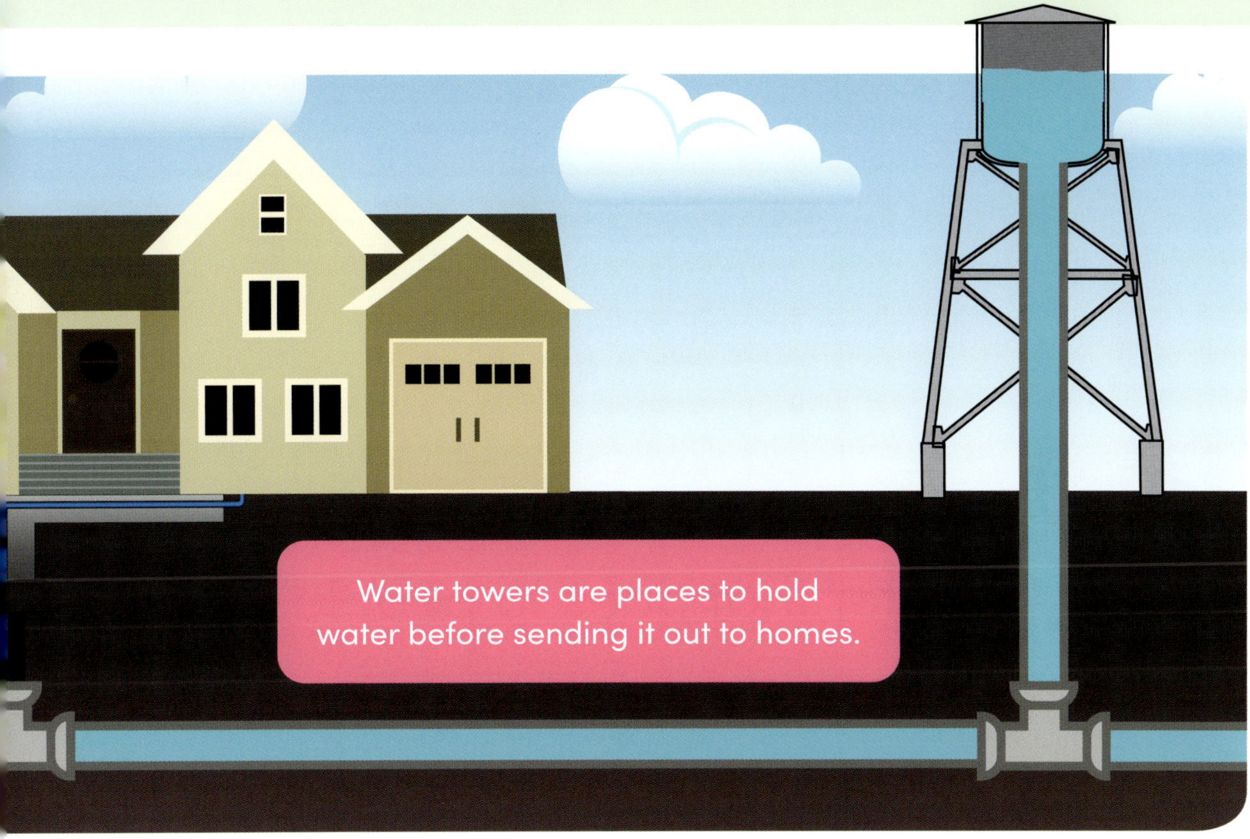

It is pumped into a water main that goes around the city. Then it is pulled into the home through pipes called lateral lines.

Chapter 4

Safe Water

All water that goes to people's homes must be filtered and tested. This keeps water safe to drink and bathe with. Safe water is called potable water. The water that goes down the drain is called wastewater.

Complete an activity here!

Cities and people have to be smart about how they use water. Sometimes an area might go through a **drought**. There will be less water available. To prepare for this, cities store large amounts of water. This is called a reservoir.

Making Connections

Text-to-Self

Were you curious about how plumbing worked? Are there any other things in your home you wished you knew more about?

Text-to-Text

Droughts are mentioned in this book. Have you read any other books that talk about droughts?

Text-to-World

What do you think the world was like before people had plumbing?

Glossary

aquifer – a layer of rock, sand, or gravel that can absorb water.

disease – a sickness.

drought – a long period of dry weather.

pressure – the pushing of a force against an opposing force.

pump – to raise or move using a pump. A pump is a tool that raises or moves liquid or gas into or out of something.

system – a set of things working together.

Index

Online Resources

popbooksonline.com

Thanks for reading this Cody Koala book!

Scan this code* and others like it in this book, or visit the website below to make this book pop!

popbooksonline.com/plumbing

*Scanning QR codes requires a web-enabled smart device with a QR code reader app and a camera.

SCIENCE QUESTIONS

WHY DO VOLCANOES ERUPT?

by Elizabeth Andrews

Cody Koala
An Imprint of Pop!
popbooksonline.com

abdobooks.com
Published by Pop!, a division of ABDO, PO Box 398166, Minneapolis, Minnesota 55439.

Printed in the United States of America, North Mankato, Minnesota

102021
012022

THIS BOOK CONTAINS RECYCLED MATERIALS

Cover Photo: Shutterstock Images
Interior Photos: Shutterstock Images, 1, 5 (top, bottom left, bottom middle), 7, 8, 11, 13, 14, 16, 19, 20–21

Editor: Grace Hansen
Series Designer: Laura Graphenteen

Library of Congress Control Number: 2021942298

Publisher's Cataloging-in-Publication Data
Names: Andrews, Elizabeth, author.
Title: Why do volcanoes erupt? / by Elizabeth Andrews
Description: Minneapolis, Minnesota : Pop!, 2022 | Series: Science questions | Includes online resources and Index.
Identifiers: ISBN 9781098241117 (lib. bdg.) | ISBN 9781098241810 (ebook)
Subjects: LCSH: Volcanoes--Juvenile literature. | Volcanic eruptions--Juvenile literature. | Magmas--Juvenile literature. | Children's questions and answers--Juvenile literature.
Classification: DDC 500--dc23

Hello! My name is

Cody Koala

Pop open this book and you'll find QR codes like this one, loaded with information, so you can learn even more!

Scan this code* and others like it while you read, or visit the website below to make this book pop.

popbooksonline.com/volcanoes-erupt

*Scanning QR codes requires a web-enabled smart device with a QR code reader app and a camera.

Table of Contents

Chapter 1

A Force of Nature

It can make the ground shake. It can make fire rain from the sky. It can create new landforms. It's a volcanic **eruption**!

Watch a video here!

Chapter 2

Earth's Layers

Planet Earth has four layers.
The crust is the rocky surface.
Below that is the mantle.
It is made of solid and
molten rock.

Learn more here!

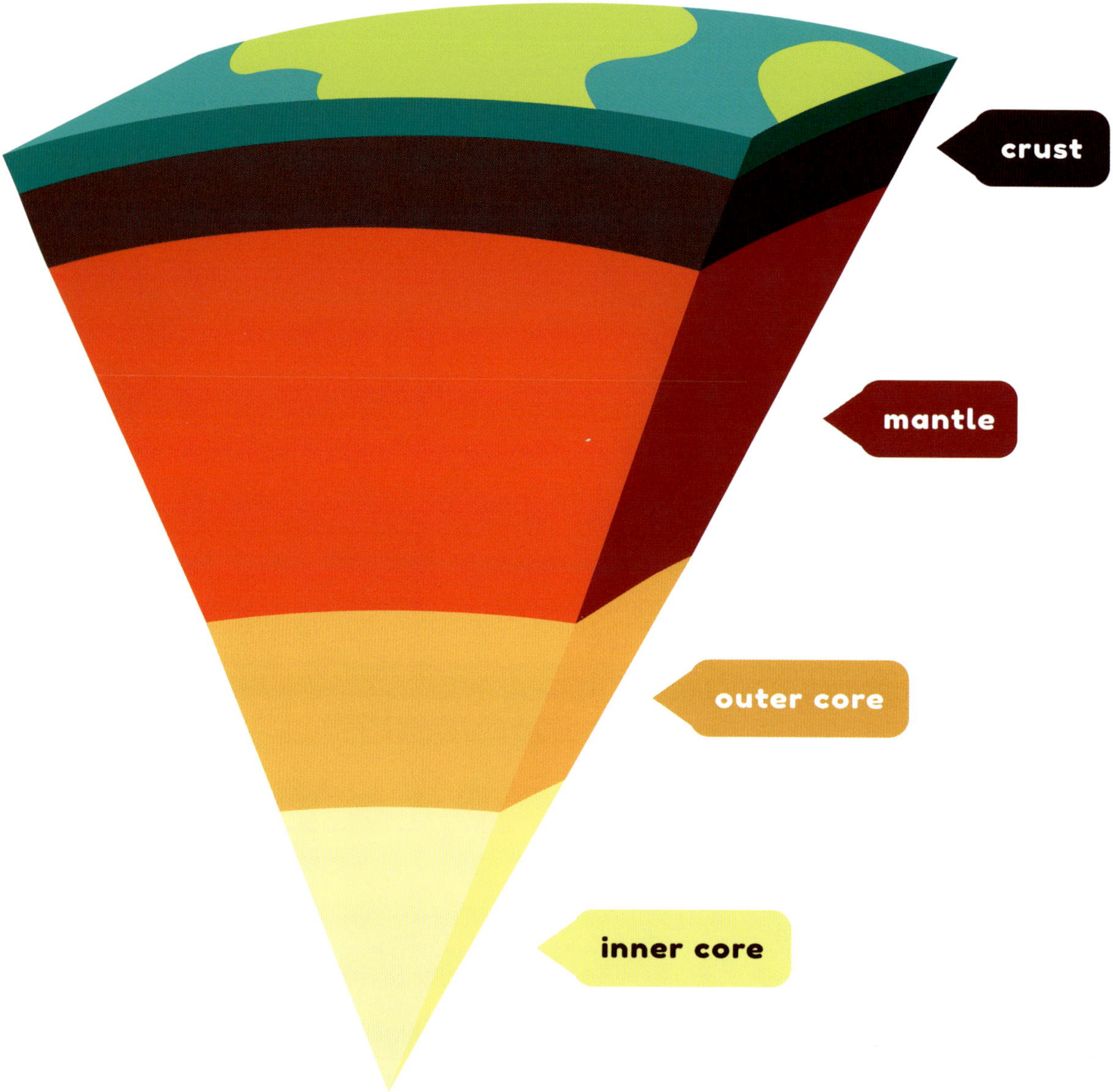
crust
mantle
outer core
inner core

Earth's **core** has an inner and an outer layer. Both are made of metals. The outer core is liquid metal. The inner core is solid metal.

The crust of Earth is split into plates of rock. They fit together like a puzzle. The plates are floating on top of the mantle's molten rock. Many volcanoes form along the edges of the plates.

When two plates meet it is called a plate boundary. You can walk along one in Iceland.

Chapter 3

What Is a Volcano?

Volcanoes are **vents** on Earth's crust. Many of them look like mountains. The opening of the volcano is called the crater. Below the crust of the volcano is a main vent.

Learn more here!

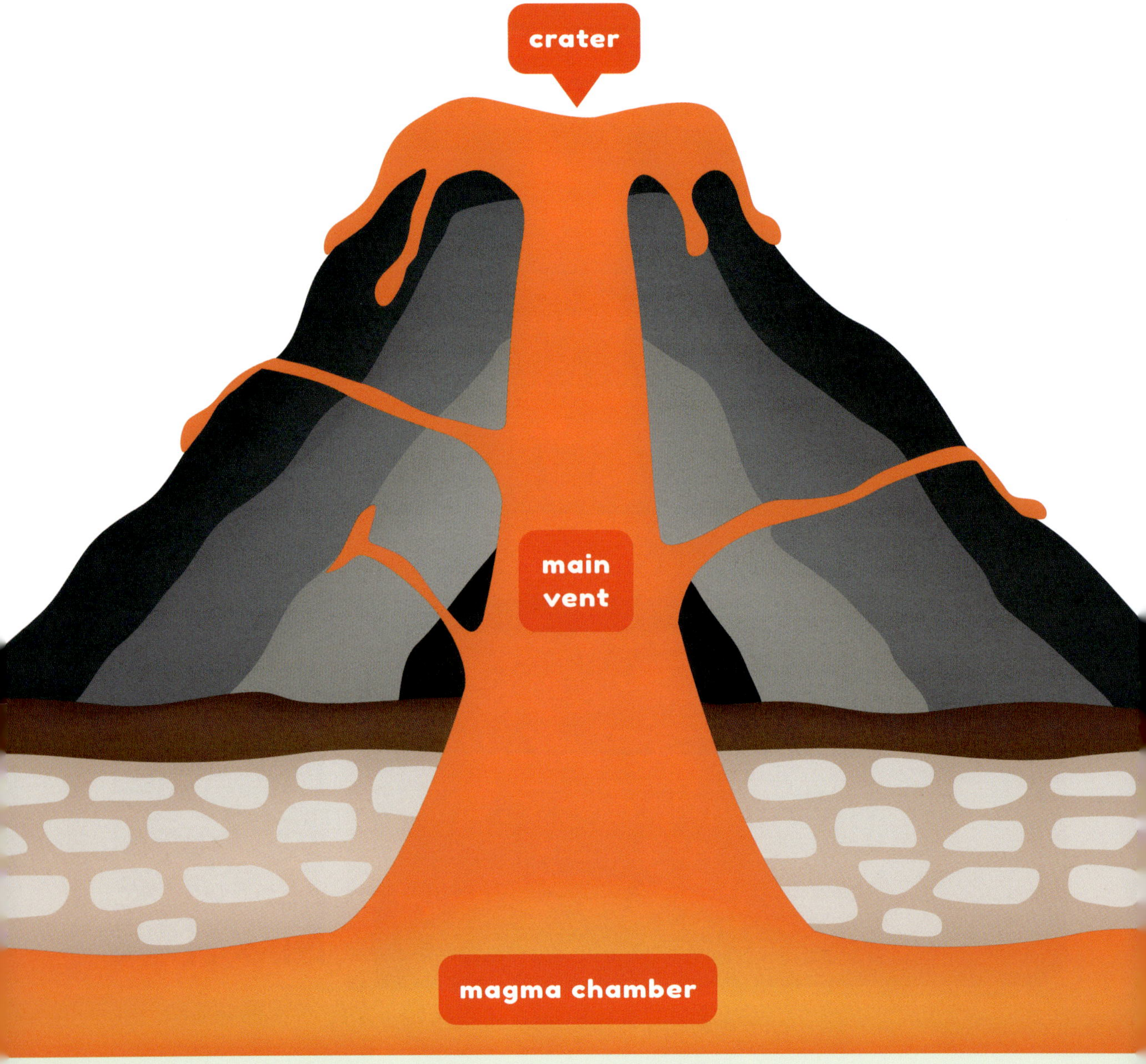
crater
main vent
magma chamber

The main vent connects a magma chamber to an opening on the crust. Magma is very hot, liquid rock below the crust. It comes from the mantle. The magma chamber is a large pool of magma that rises to the surface.

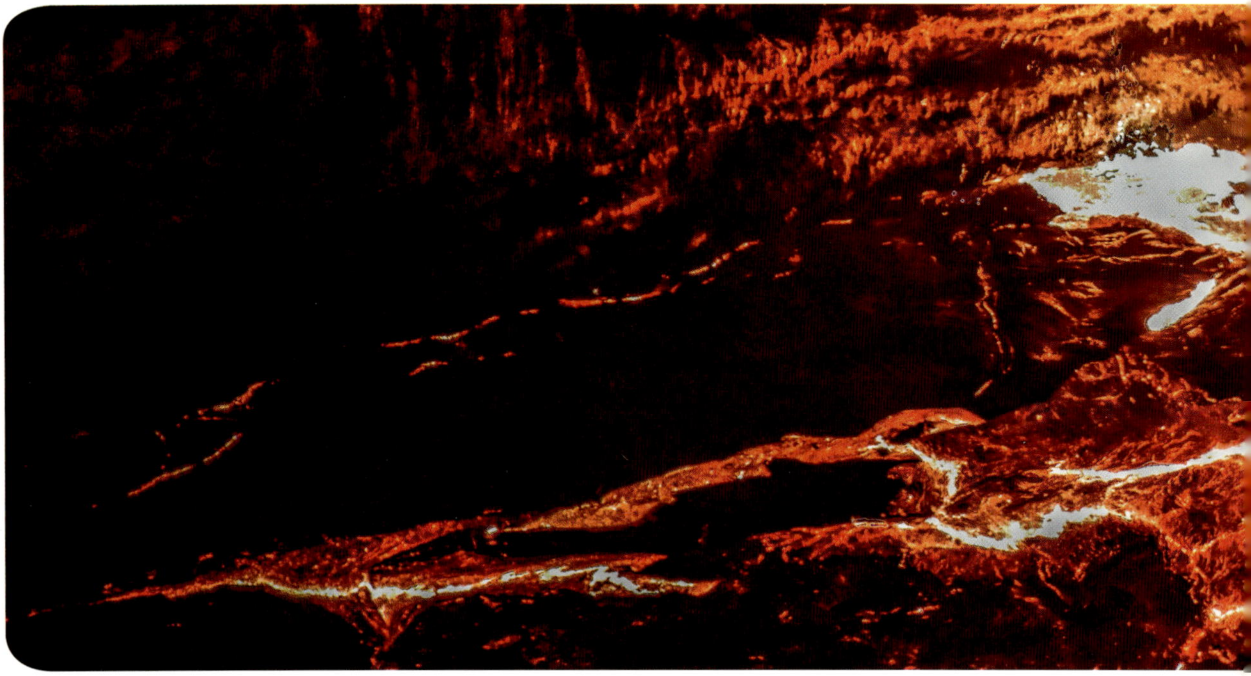

Magma can rise because it is less **dense** than solid rock. It flows through cracks in the rocks of the crust to

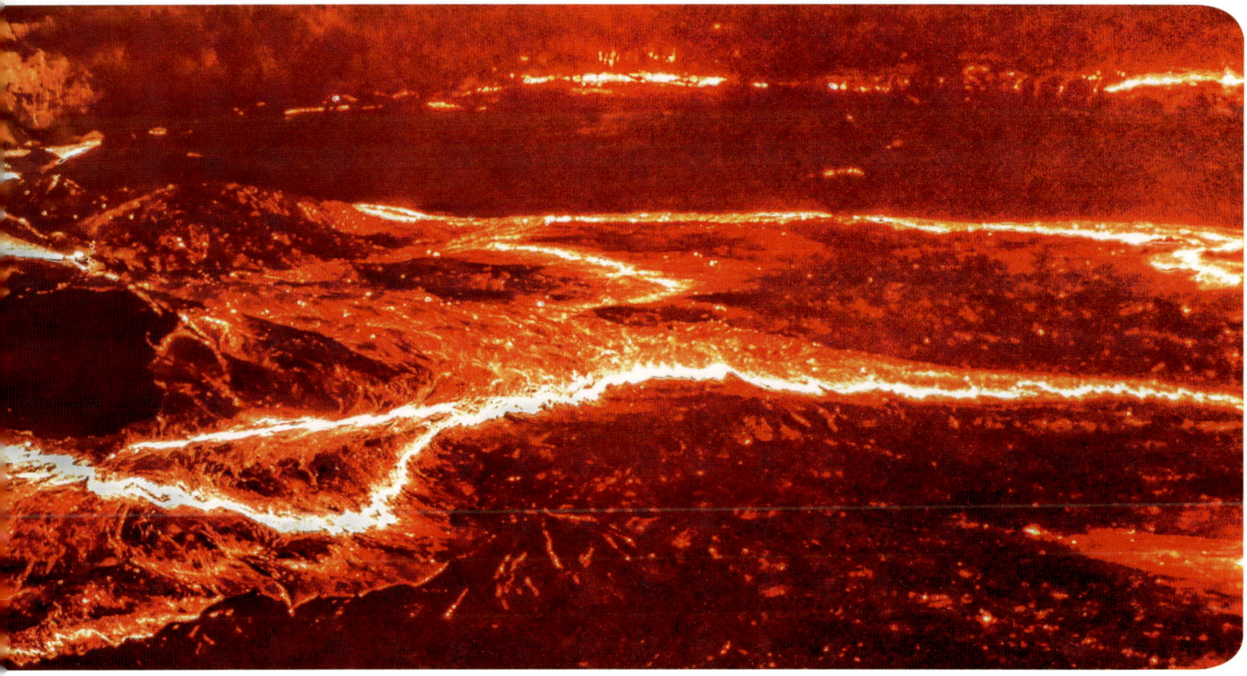

collect in the chamber. The chamber also holds gas created by the magma.

Chapter 4

Eruption!

As magma rises, the **pressure** in the chamber builds. All the gas and **molten** rock filling the space needs a place to go. It will be pushed up and out through the **vent**. This is an **eruption**.

When magma reaches the surface it is called lava.

Complete an activity here!

Eruptions usually last for two months. Sometimes the eruption is explosive. Gas, lava, and pieces of rock will

shoot into the air. An eruption can also be calm. Lava might flow gently from the volcano's crater.

Making Connections

Text-to-Self

Would you ever want to see a volcano erupt? How would you stay safe while watching?

Text-to-Text

Have you read any other books about natural disasters? How were they similar to and different from a volcanic eruption?

Text-to-World

Do you think it's a good idea for people to live near volcanoes? Explain why or why not.

Glossary

core – the very hot, dense center of Earth.

dense – tightly packed.

erupt – to force out or release, sometimes suddenly.

molten – made liquid by very high heat.

pressure – a steady force upon a surface.

vent – an opening that allows gas or liquid to pass through.

Index

Online Resources

popbooksonline.com

Thanks for reading this Cody Koala book!

Scan this code* and others like it in this book, or visit the website below to make this book pop!

popbooksonline.com/volcanoes-erupt

*Scanning QR codes requires a web-enabled smart device with a QR code reader app and a camera.

SCIENCE QUESTIONS

WHY DOES A BOAT FLOAT?

by Elizabeth Andrews

Cody Koala
An Imprint of Pop!
popbooksonline.com

abdobooks.com
Published by Pop!, a division of ABDO, PO Box 398166, Minneapolis, Minnesota 55439.

Printed in the United States of America, North Mankato, Minnesota

102021
012022
THIS BOOK CONTAINS RECYCLED MATERIALS

Cover Photo: Shutterstock Images
Interior Photos: Shutterstock Images, 1, 5 (top, bottom left, bottom center), 7, 8–9, 11, 12 (top, bottom), 15, 17 (top, bottom left, bottom center), 18–19, 21

Editor: Grace Hansen
Series Designer: Laura Graphenteen

Library of Congress Control Number: 2021942249

Publisher's Cataloging-in-Publication Data
Names: Andrews, Elizabeth, author.
Title: Why does a boat float? / by Elizabeth Andrews
Description: Minneapolis, Minnesota : Pop!, 2022 | Series: Science questions | Includes online resources and index.
Identifiers: ISBN 9781098241124 (lib. bdg.) | ISBN 9781098241827 (ebook)
Subjects: LCSH: Boats--Juvenile literature. | Buoyant ascent (Hydrodynamics)--Juvenile literature. | Density--Juvenile literature. | Children's questions and answers--Juvenile literature.
Classification: DDC 500--dc23

Hello! My name is

Cody Koala

Pop open this book and you'll find QR codes like this one, loaded with information, so you can learn even more!

Scan this code* and others like it while you read, or visit the website below to make this book pop.
popbooksonline.com/boat-float

*Scanning QR codes requires a web-enabled smart device with a QR code reader app and a camera.

Table of Contents

Chapter 1

What Floats That Boat?

Have you ever wondered why something as big as a cruise ship is able to stay on top of the water? There are even boats big enough for airplanes to take off from and land on. How do these boats float?

The oldest boat ever discovered is nearly 10,000 years old.

Watch a video here!

Chapter 2

Buoyancy

Boats can float because they are **buoyant**. Things are buoyant because different forces act on each other. The upward force of the water pushes against the downward force of the boat.

Learn more here!

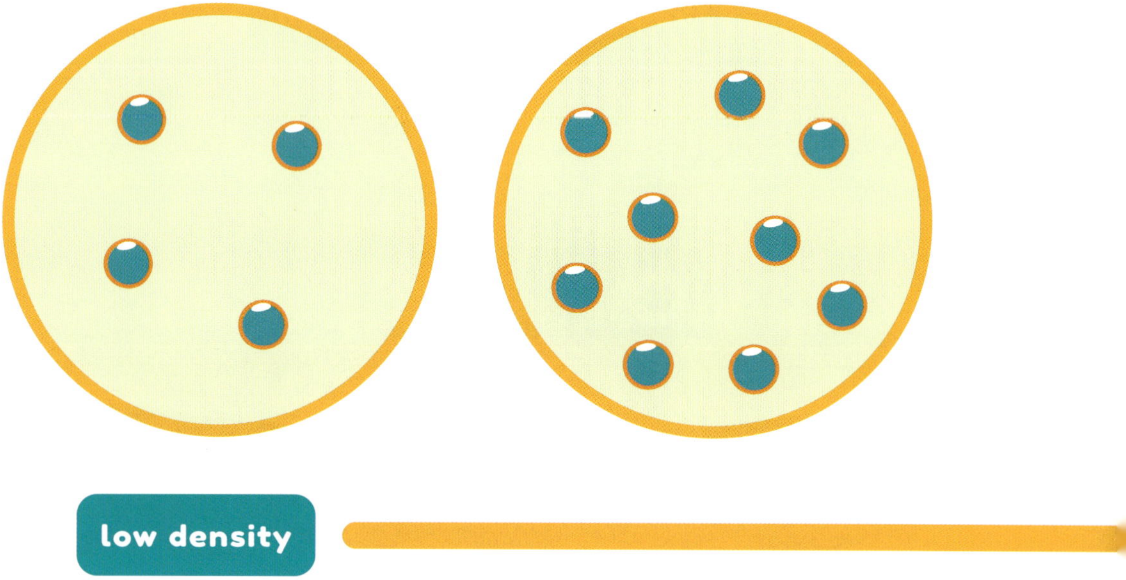

These forces are **determined** by the density of the boat and the density of the water.

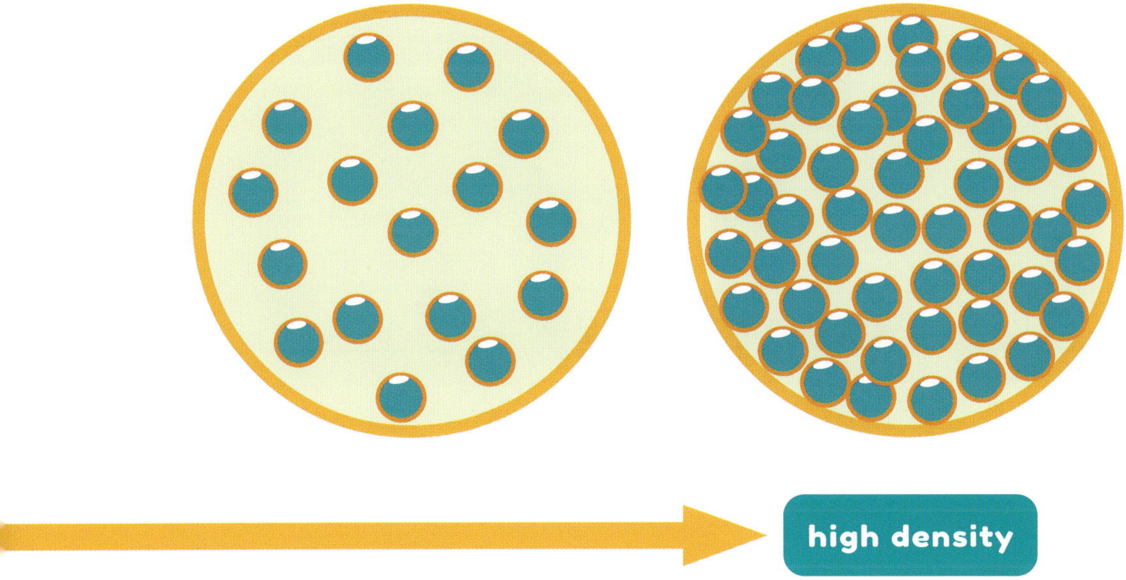

Density is how much space an object takes up compared to how much matter is in the object.

Chapter 3

What Matters?

Matter is anything that takes up space. Everything is made up of matter. The air you breathe, the apple you eat, and even *you* have matter. Mass measures the amount of matter something has.

Learn more here!

A sponge takes up more space than a rock. But it does not have more mass than the rock. This is because the rock is denser. It has more matter inside a smaller space than the sponge.

Even though a boat is big, it doesn't have a lot of matter inside to give it a large mass. There is air in the boat. Air has low density. This keeps the boat's density low.

An inner tube keeps you afloat because it is filled with air. When you're in the tube, your mass spreads across the tube. This lowers your density.

Chapter 4

Above the Surface

When a boat enters water, it's density **displaces** some of the water. The amount of water that moved is equal to the amount of space the boat takes up.

Complete an activity here!

The shape of the boat is also important. A boat has a wide **hull** so its mass can be spread out. The larger the hull is, the more water will be displaced.

This means there will be more force pushing up on the boat.

The density of the boat pushes down against the water. But the water also pushes up on the boat. As long as the boat's mass is less than the mass of the water it displaces, it will stay afloat!

You should always wear a life jacket when you're aboard a boat.

Making Connections

Text-to-Self

Have you ever been on a boat? If so, how big was it? If not, what kind of boat would you first want to go on?

Text-to-Text

Have you read any other books that discuss density and mass? How were they similar to and different from this book?

Text-to-World

How do you think boats changed the world? Write a few sentences explaining your answer.

Glossary

buoyant – having the ability to float.

determine – to decide.

displace – to physically move out of position.

hull – the main part of a boat or ship including the deck, sides, and bottom.

Index

Online Resources

popbooksonline.com

Thanks for reading this Cody Koala book!

Scan this code* and others like it in this book, or visit the website below to make this book pop!

popbooksonline.com/boat-float

*Scanning QR codes requires a web-enabled smart device with a QR code reader app and a camera.

SCIENCE QUESTIONS

HOW DOES A BODY HEAL?

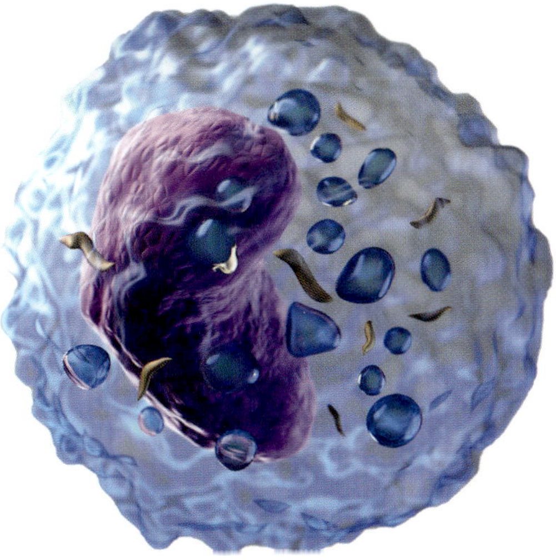

by Elizabeth Andrews

Cody Koala
An Imprint of Pop!
popbooksonline.com

abdobooks.com
Published by Pop!, a division of ABDO, PO Box 398166, Minneapolis, Minnesota 55439.

Printed in the United States of America, North Mankato, Minnesota

102021
012022

THIS BOOK CONTAINS RECYCLED MATERIALS

Cover Photo: Shutterstock Images
Interior Photos: Getty Images, 8, 10, 13, 18, 19; Shutterstock Images, 5, 7, 9, 14, 17, 21

Editor: Grace Hansen
Series Designer: Laura Graphenteen

Library of Congress Control Number: 2021942433

Publisher's Cataloging-in-Publication Data
Names: Andrews, Elizabeth, author.
Title: How does a body heal? / by Elizabeth Andrews
Description: Minneapolis, Minnesota : Pop!, 2022 | Series: Science questions | Includes online resources and index.
Identifiers: ISBN 9781098241094 (lib. bdg.) | ISBN 9781098241797 (ebook)
Subjects: LCSH: Human body--Juvenile literature. | Healing--Juvenile literature. | Medicine--Juvenile literature. | Children's questions and answers--Juvenile literature.
Classification: DDC 500--dc23

Hello! My name is

Cody Koala

Pop open this book and you'll find QR codes like this one, loaded with information, so you can learn even more!

Scan this code* and others like it while you read, or visit the website below to make this book pop.

popbooksonline.com/body-heal

*Scanning QR codes requires a web-enabled smart device with a QR code reader app and a camera.

Table of Contents

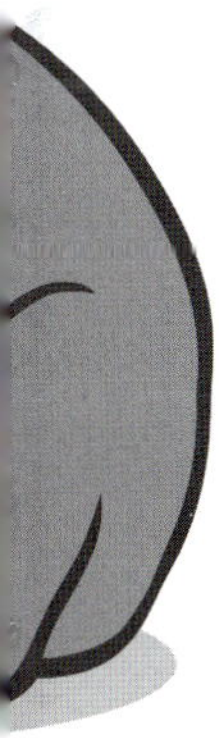

Chapter 1

Tiny Workers

Everybody is made up of cells. A cell is a tiny living thing. About 200 different kinds of cells make up the human body. They all have a different job to do, like helping you heal!

Watch a video here!

Chapter 2

Healing a Cut

When you scrape your knee, your body will send out an alert that it's time to start healing. The alerts are **inflammatory responses**. These also keep **germs** from spreading farther into the body.

If too many germs enter through the cut, you might get a fever. The high temperature helps kill more germs.

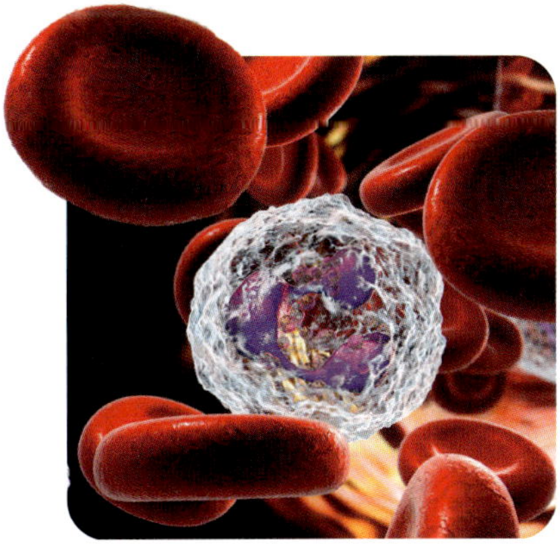

Learn more here!

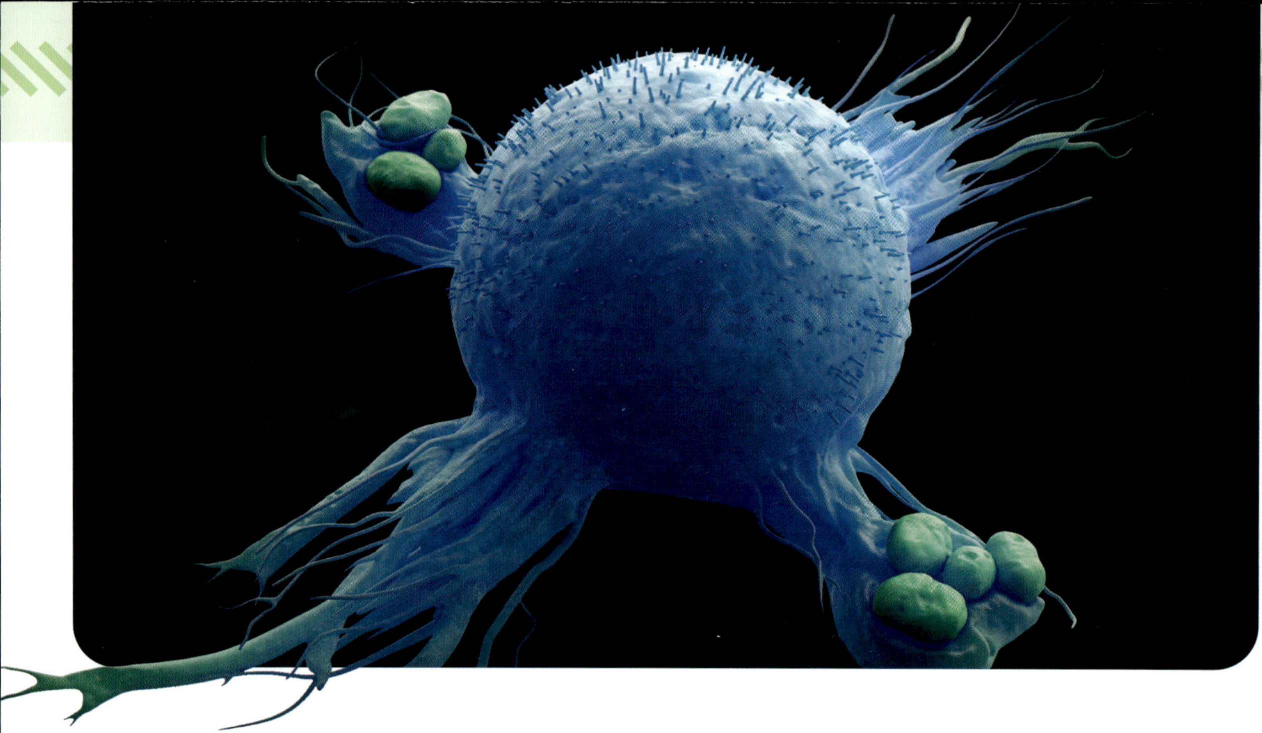

The **immune system** gets to work right away to protect you from infection. Cells called phagocytes eat germs and dirt getting in through the open skin.

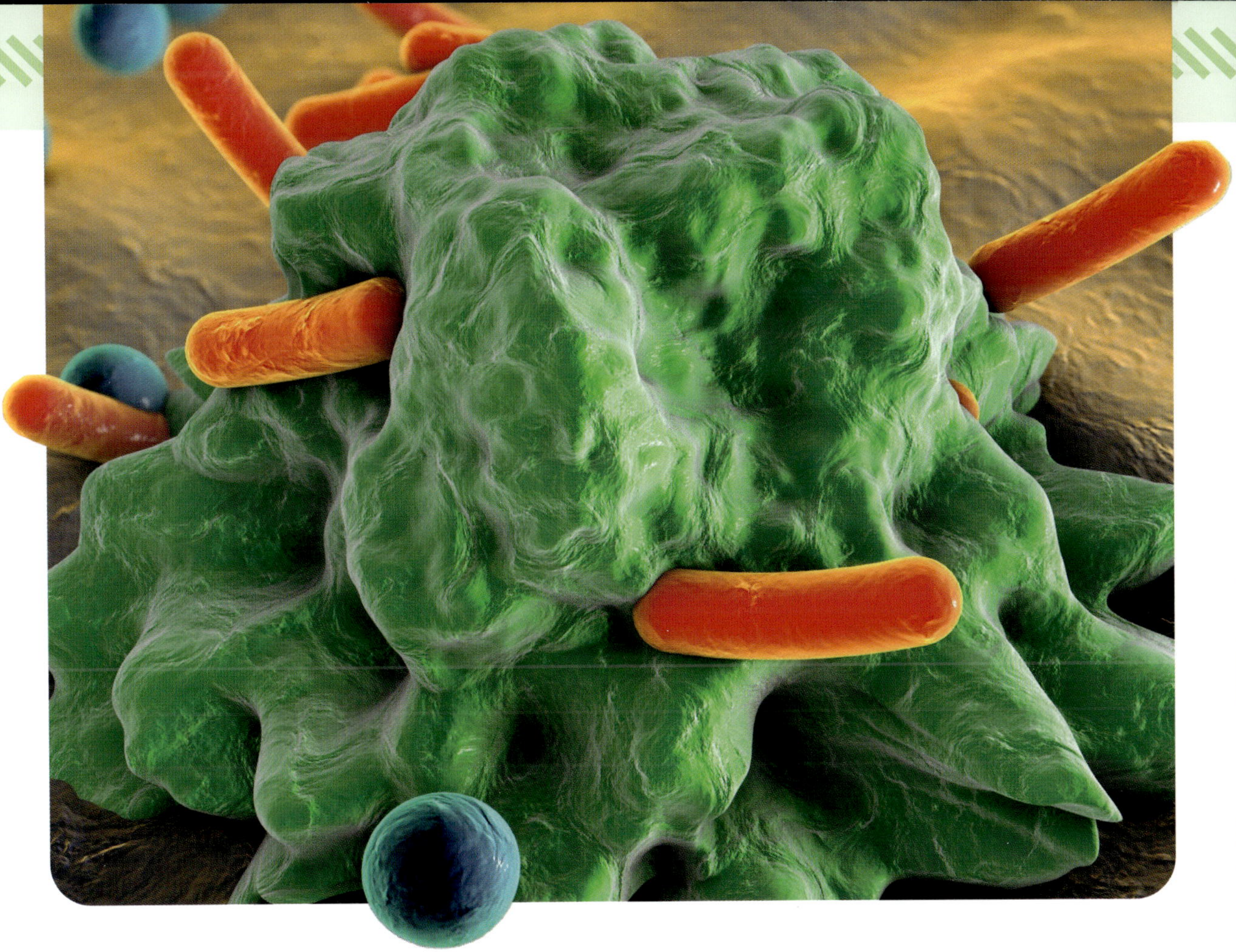

Phagocytes will also eat cells around the scrape that are too damaged to heal.

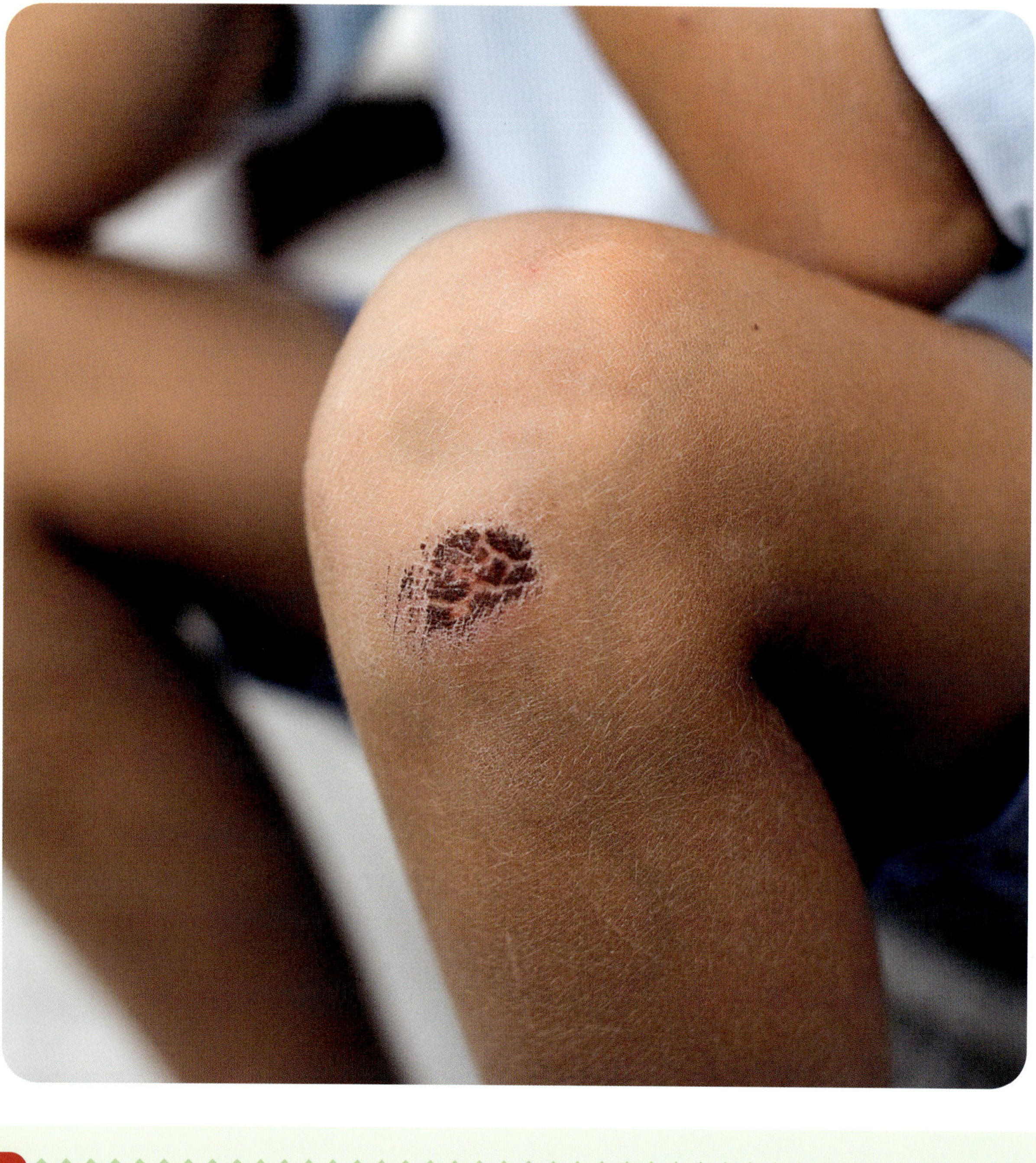

Blood cells clump together in the scrape. The blood hardens to form a scab. Now there is a natural **barrier** to stop any new germs from getting in. More cells come to help build new **tissue**.

Chapter 3

Healing a Broken Bone

A broken bone is more serious than a scrape. But the healing happens much the same. When the bone breaks, its cells send out signals. **Inflammatory responses** begin.

Learn more here!

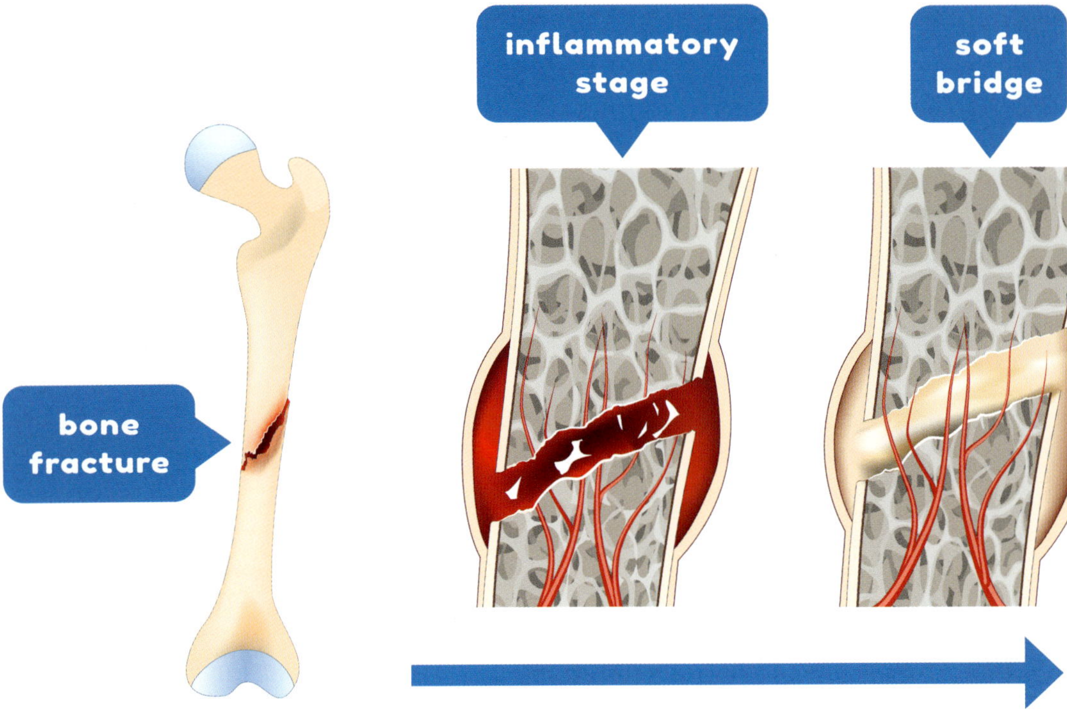

There will be swelling around the break. Blood and bone cells work to form a soft bridge between the pieces of bone.

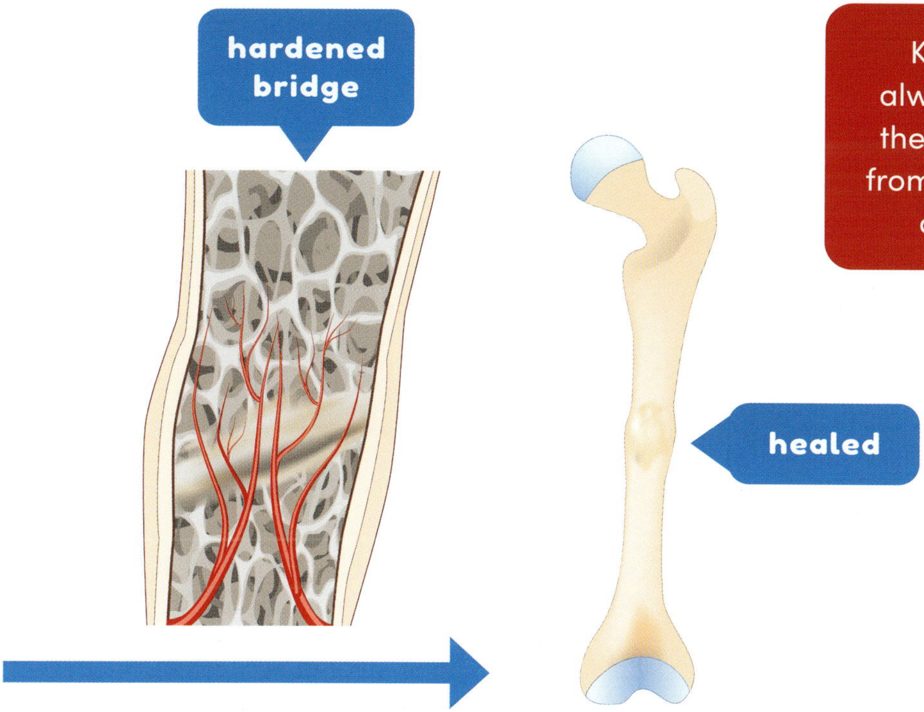

Kids' bones are always growing. So they will heal faster from a break than an adult's would.

Then osteoblast cells rebuild the bone. They bring calcium and **minerals** to the bridge. This will build new, strong bone.

Chapter 4

Healing from Sickness

Your body will notice if bad **germs** invade it. This invasion needs help from your **immune system**. It begins the healing with **inflammatory responses**.

Complete an activity here!

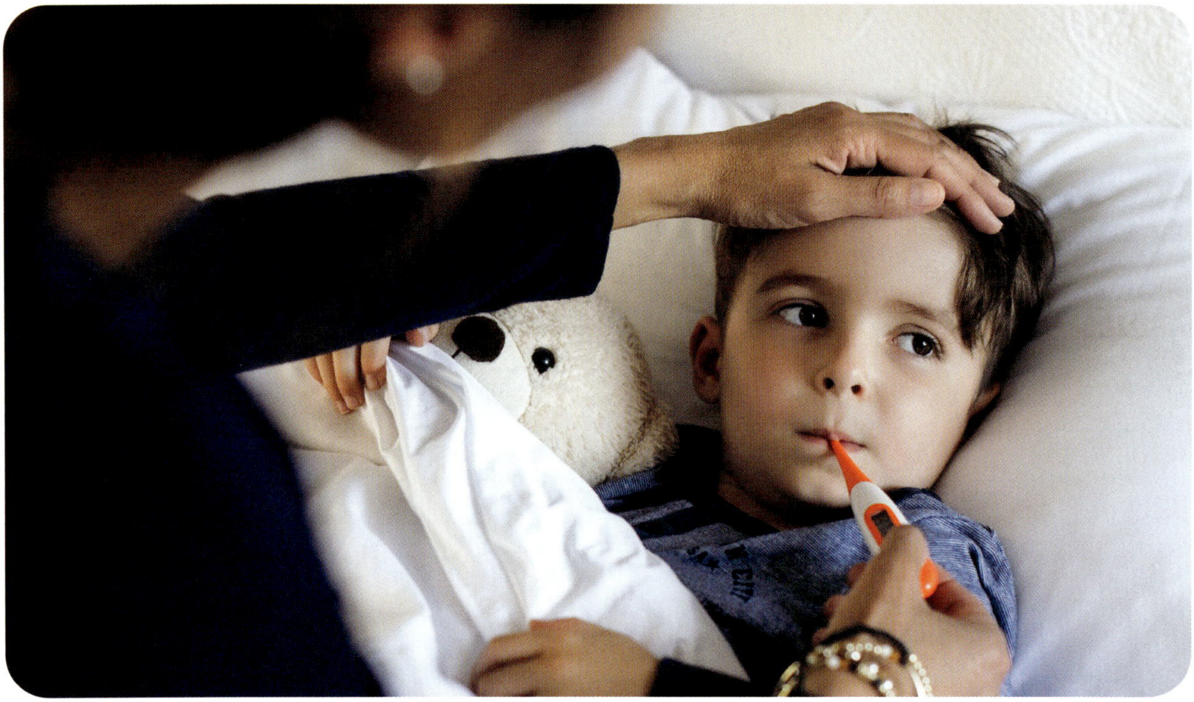

A fever will start to burn away any infection. The immune system's B-cells and **antibodies** mark invaders.

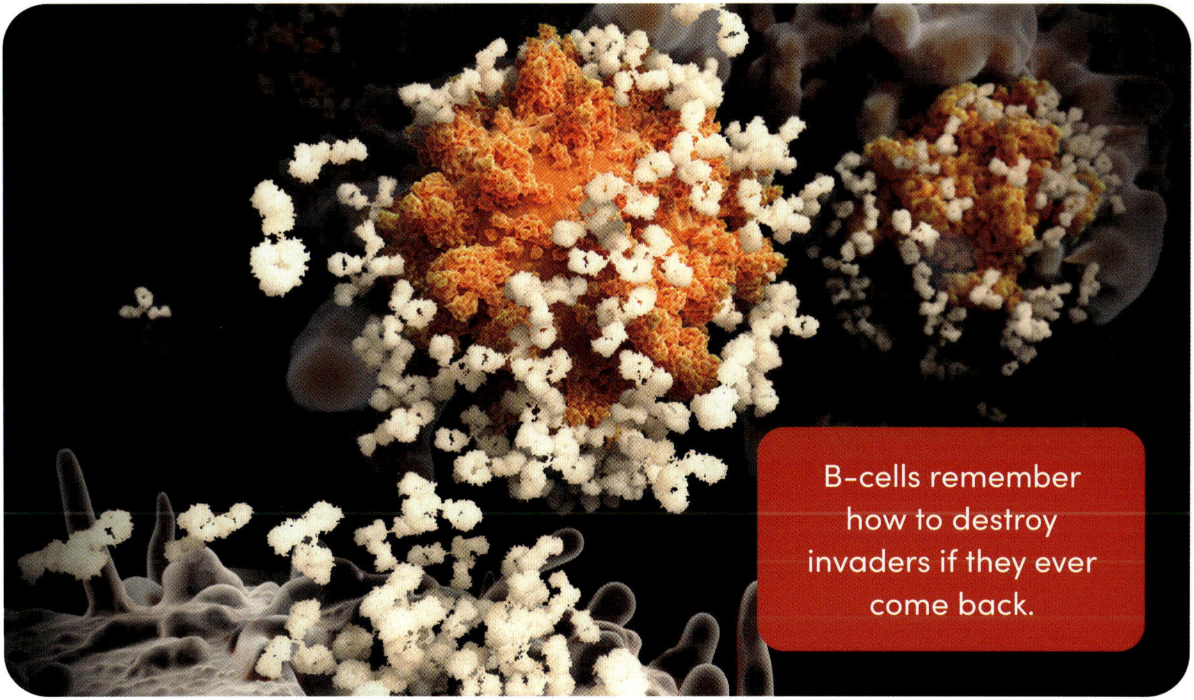
B-cells remember how to destroy invaders if they ever come back.

Now, phagocyte cells can easily find and destroy them! Soon you will start feeling better!

By eating healthy foods and exercising, you can strengthen your immune system. This will help make sure that your body is always ready to heal. You should also get enough sleep so your cells can mend while you rest.

Making Connections

Text-to-Self

Have you ever healed from an injury? How long did it take?

Text-to-Text

Have you read any other books about your body before? Did you learn anything different in them?

Text-to-World

Are there people or groups who can help with healing? What are they like?

Glossary

antibody – a substance produced by special cells of the body to fight sickness.

barrier – something that blocks.

germs – tiny organisms that cause sickness.

immune system – the system that protects the body from sickness.

inflammatory response – the way a body reacts to injury or sickness. Kinds of responses are redness, swelling, and fever.

mineral – a nonliving thing that comes from the earth and food. The body needs minerals to work.

tissue – a group of the same cells.

Index

Online Resources

popbooksonline.com

Thanks for reading this Cody Koala book!

Scan this code* and others like it in this book, or visit the website below to make this book pop!

popbooksonline.com/body-heal

*Scanning QR codes requires a web-enabled smart device with a QR code reader app and a camera.

SCIENCE QUESTIONS

HOW DO FISH BREATHE?

by Elizabeth Andrews

Cody Koala
An Imprint of Pop!
popbooksonline.com

abdobooks.com
Published by Pop!, a division of ABDO, PO Box 398166, Minneapolis, Minnesota 55439.

Printed in the United States of America, North Mankato, Minnesota

102021
012022

THIS BOOK CONTAINS RECYCLED MATERIALS

Cover Photo: Shutterstock Images
Interior Photos: Getty Images, 6; Shutterstock Images 5, 9 10, 13, 15, 16, 18, 21

Editor: Grace Hansen
Series Designer: Laura Graphenteen

Library of Congress Control Number: 2021942247

Publisher's Cataloging-in-Publication Data
Names: Andrews, Elizabeth, author.
Title: How do fish breathe? / by Elizabeth Andrews
Description: Minneapolis, Minnesota : Pop!, 2022 | Series: Science questions | Includes online resources and index
Identifiers: ISBN 9781098241070 (lib. bdg.) | ISBN 9781098241773 (ebook)
Subjects: LCSH: Fishes--Juvenile literature. | Fishes--Respiration--Juvenile literature. | Gills--Juvenile literature. | Fishes--Behavior--Juvenile literature. | Children's questions and answers--Juvenile literature.
Classification: DDC 500--dc23

Hello! My name is

Cody Koala

Pop open this book and you'll find QR codes like this one, loaded with information, so you can learn even more!

Scan this code* and others like it while you read, or visit the website below to make this book pop.

popbooksonline.com/fish-breathe

*Scanning QR codes requires a web-enabled smart device with a QR code reader app and a camera.

Table of Contents

Chapter 1

Something in the Water

Fish need **oxygen** to live, just like humans and land animals. But fish live underwater. They can't get oxygen from the air. Instead, fish get oxygen from water.

Watch a video here!

Fast moving, cold water holds the most oxygen.

Water and air are made up of tiny things called **molecules**. There are different kinds of molecules in water.

Learn more here!

gills

Gills need to be protected. Strong but flexible bony plates called operculum lay over the gills. The capillaries are also supported by bony loops called gill arches.

Fish gills open and close to push water out after the **oxygen** has been taken in.

As the blood flows through the capillaries, it takes in the oxygen **molecules** from the water. The oxygen attaches to the blood. Then it is carried through the fish's body to its **cells**.

BREATHING WATER

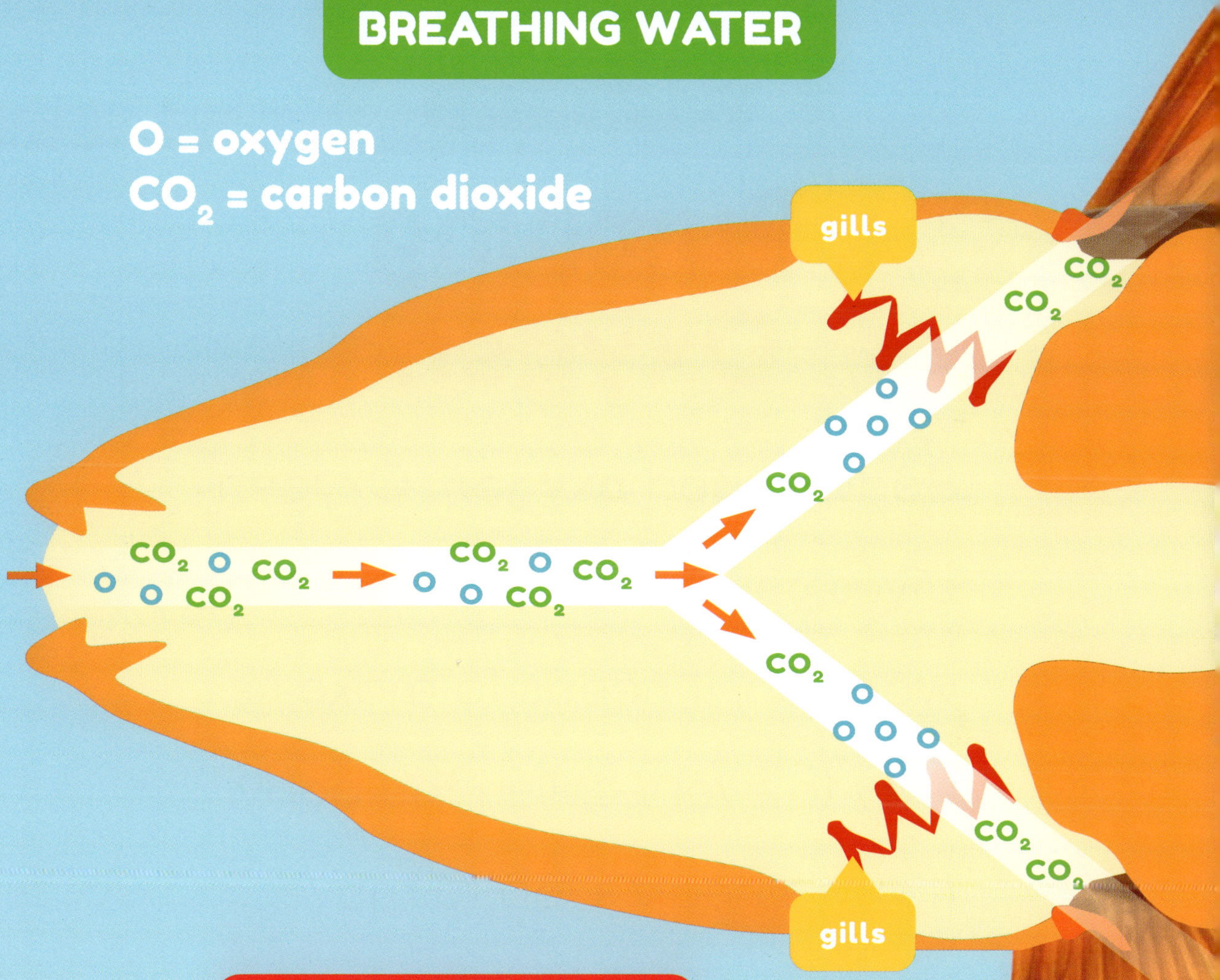

Fish breathe out something called carbon dioxide. Plants in the water will use it.

Chapter 3

Pumping Oxygen

Bodies need **oxygen**. Oxygen gives **cells** what they need to work. So, it needs to make its way to every cell in the body. It is the heart's job to move the oxygen-filled blood.

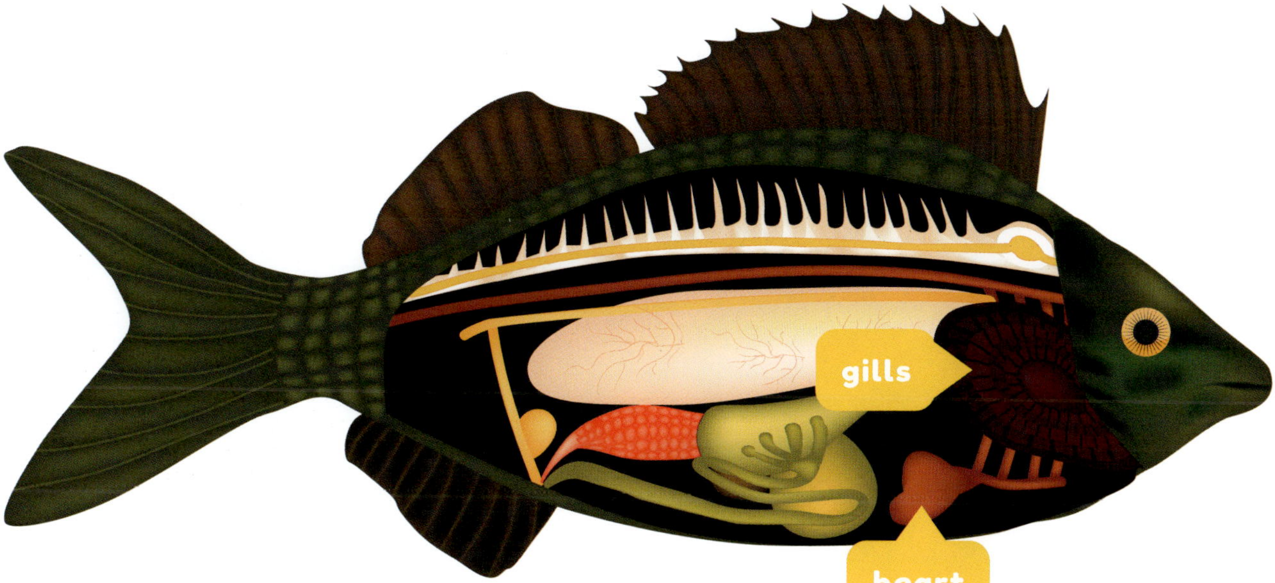

Learn more here!

Water is thicker than air.
It also has less oxygen in it
than air.

This means that it takes more work for a fish to get the oxygen it needs.

capillaries

Since water is low in oxygen, fish had to adapt. The capillaries in the gills have a lot of little folds. These give fish more chances to take in the oxygen in the water.

Chapter 4

Changing Waters

Scientists watch **oxygen** levels in water. They noticed that levels in the ocean are decreasing. **Climate change** might be the cause of this. Scientists are working on ways to slow down the fall of oxygen levels.

Complete an activity here!

Making Connections

Text-to-Self

Have you ever seen fish gills up close? Where were you when you saw them?

Text-to-Text

Have you read any other books about fish? Did you learn anything that wasn't in this book?

Text-to-World

Fish breathe with their gills. What parts of the body help humans breathe?

Glossary

cell – the simplest unit of life that makes up all living things.

climate change – a long-term change in Earth's climate. It includes changing temperatures and weather patterns. It can result from natural processes or human activities.

current – a flow of water.

dissolve – to become liquid.

molecule – the smallest part of a material.

oxygen – a colorless, tasteless gas in the air. It is needed to live.

vessel – a tube that blood moves through.

Index

Online Resources

popbooksonline.com

Thanks for reading this Cody Koala book!

Scan this code* and others like it in this book, or visit the website below to make this book pop!

popbooksonline.com/fish-breathe

*Scanning QR codes requires a web-enabled smart device with a QR code reader app and a camera.

SCIENCE QUESTIONS

HOW DO PUPPIES GET THEIR COLORS?

by Elizabeth Andrews

Cody Koala
An Imprint of Pop!
popbooksonline.com

abdobooks.com
Published by Pop!, a division of ABDO, PO Box 398166, Minneapolis, Minnesota 55439.

Printed in the United States of America, North Mankato, Minnesota

102021
012022
THIS BOOK CONTAINS RECYCLED MATERIALS

Cover Photo: Shutterstock Images
Interior Photos: Shutterstock Images, 1, 5 (top, bottom), 6, 7, 9, 10, 13, 14, 17, 18, 21 (top, bottom left), Wikimedia, 17 (bottom middle)

Editor: Grace Hansen
Series Designer: Laura Graphenteen

Library of Congress Control Number: 2021942296

Publisher's Cataloging-in-Publication Data
Names: Andrews, Elizabeth, author.
Title: How do puppies get their colors? / by Elizabeth Andrews
Description: Minneapolis, Minnesota : Pop!, 2022 | Series: Science questions | Includes online resources and index.
Identifiers: ISBN 9781098241087 (lib. bdg.) | ISBN 9781098241780 (ebook)
Subjects: LCSH: Puppies--Juvenile literature. | Animals--Color--Juvenile literature. | Puppies--Breeding--Juvenile literature. | Children's questions and answers--Juvenile literature.
Classification: DDC 500--dc23

Hello! My name is

Cody Koala

Pop open this book and you'll find QR codes like this one, loaded with information, so you can learn even more!

Scan this code* and others like it while you read, or visit the website below to make this book pop.

popbooksonline.com/puppy-colors

*Scanning QR codes requires a web-enabled smart device with a QR code reader app and a camera.

Table of Contents

Chapter 1

Pick of the Litter

There is nothing quite as cute as a litter of puppies. A litter is the group of puppies that a female dog has at one time.

Watch a video here!

Picture a litter of dachshunds (dahks-hunds). Most of the young pups have black coats, like their parents. But there are two that are brown.

Puppies from the same litter can look different from one another. Genes are the reason for these differences.

Litters can have anywhere from 1–12 puppies in them!

Chapter 2

What Are Genes?

Genes are instructions that tell the body how to look and act. Every person and animal has two copies of each gene, one from each parent. Genes are made of **DNA**.

Learn more here!

If all DNA inside a single chromosome was unwrapped, it would stretch 6 feet (1.8 m)!

Genes are found in the middle of **cells**, on chromosomes. Chromosomes are tightly packed clumps of DNA. People often have 23 pairs of chromosomes in each cell. Most of them are shaped like an X.

Chapter 3

What Are Alleles?

The **combination** of genes a puppy gets from its parents decides the puppy's physical **traits**. Every puppy has two copies of each gene. Alleles are forms of the same gene.

Learn more here!

Capital and lowercase letters stand for alleles. Think of the dachshund litter again. Both parents have black coats. They also each have *Bb* alleles. Capital *B* is for black. Lowercase *b* is for brown. This means each parent could pass either trait on to each puppy.

Chapter 4

Punnett Square

Reginald Punnett created a simple way to predict how **traits** may get passed on. It is called the Punnett square. Each section in the square shows an allele that a puppy could get.

Complete an activity here!

Reginald Punnett

DACHSHUND PUPPY COAT COLOR

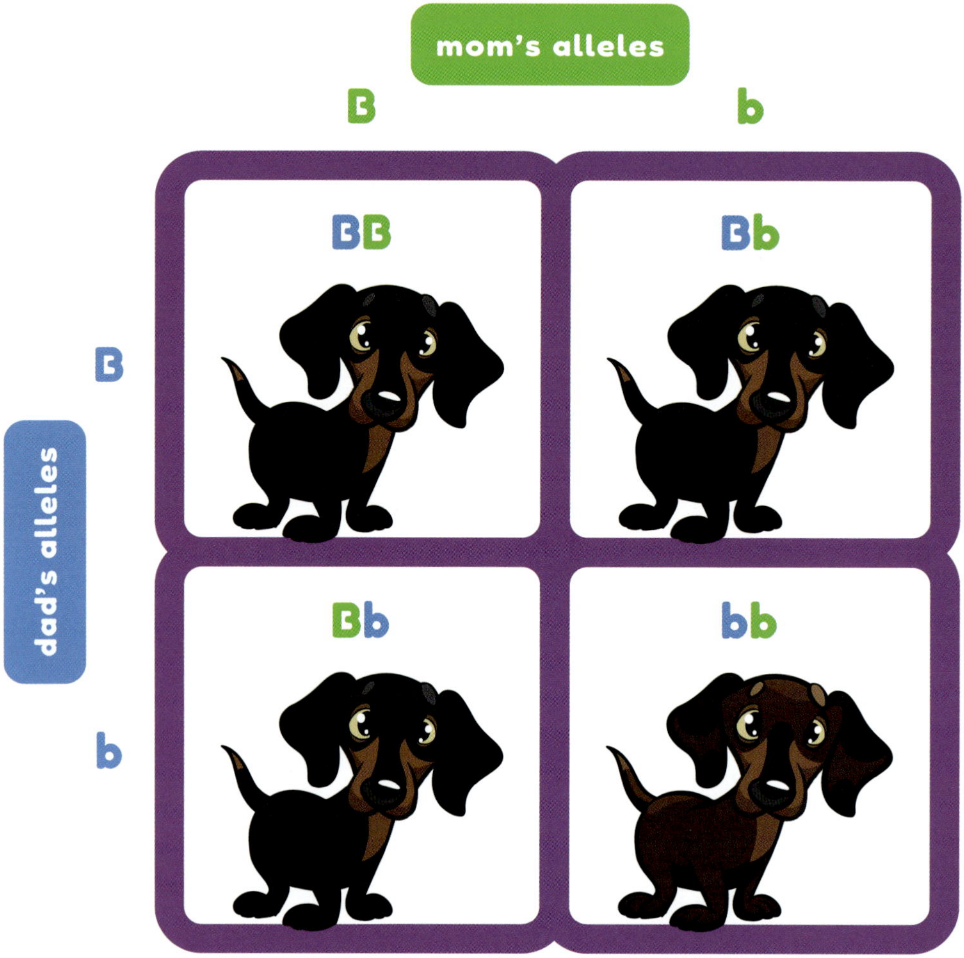

UPPERCASE LETTER: **dominant** gene
lowercase letter: **recessive** gene

If a puppy **inherits** even one uppercase letter, it will have a black coat. In this litter, one *bb* **combination** is possible. This means the puppies only have a one-in-four chance to have a brown coat. They are more likely to have black coats.

Humans have genes too. They inherit different ones from each parent just like puppies. This is why people don't all look the same. Everyone is made of a special combination of genes!

Identical twins look exactly the same because they inherited matching genes!

Making Connections

Text-to-Self

Have you ever seen a litter of puppies? Did they look alike or different from one another?

Text-to-Text

Have you read any other books about genes? If so, did you learn anything different in that book?

Text-to-World

Do you think it's good that everyone looks different? What would the world be like if people all looked the same?

Glossary

cell – the smallest part of a living thing.

combination – a mixture of different things.

DNA – short for deoxyribonucleic acid. A material in the body that helps determine what features a living thing will have.

dominant – relating to the controlling gene in an allele.

inherit – to receive from a parent.

recessive – relating to a gene that will not be expressed if paired with a dominant gene in an allele.

trait – a quality or feature of something.

Index

Online Resources

popbooksonline.com

Thanks for reading this Cody Koala book!

Scan this code* and others like it in this book, or visit the website below to make this book pop!

popbooksonline.com/puppy-colors

*Scanning QR codes requires a web-enabled smart device with a QR code reader app and a camera.